T0183454

Lecture Notes
in Business Information Processing

425

More information about this series at http://www.springer.com/series/7911

Inès Saad · Camille Rosenthal-Sabroux ·
Faiez Gargouri · Pierre-Emmanuel Arduin (Eds.)

Information and Knowledge Systems

Digital Technologies, Artificial Intelligence and Decision Making

5th International Conference, ICIKS 2021
Virtual Event, June 22–23, 2021
Proceedings

 Springer

Editors
Inès Saad (iD)
Amiens Business School and University
of Picardie Jules Verne
Amiens, France

Camille Rosenthal-Sabroux
University of Paris-Dauphine PSL
Paris, France

Faiez Gargouri
University of Sfax
Sfax, Tunisia

Pierre-Emmanuel Arduin
University of Paris-Dauphine PSL
Paris, France

ISSN 1865-1348   ISSN 1865-1356   (electronic)
Lecture Notes in Business Information Processing
ISBN 978-3-030-85976-3   ISBN 978-3-030-85977-0   (eBook)
https://doi.org/10.1007/978-3-030-85977-0

This Springer imprint is published by the registered company Springer Nature Switzerland AG
The registered company address is: Gewerbestrasse 11, 6330 Cham, Switzerland

Preface

We are pleased to present the proceedings of the fifth edition of the International Conference on Information and Knowledge Systems (ICIKS 2021, previously named KMIKS), held online during June 22–23, 2021.

The International Conference on Information and Knowledge Systems (ICIKS 2021) gathered both researchers and practitioners in the fields of information systems, artificial intelligence, knowledge management and decision support. ICIKS seeks to promote discussions on various organizational, technological, and socio-cultural aspects of research in the design and use of information and knowledge systems in organizations.

Considering the large amount of data, information, and knowledge created and used in organizations, especially with the evolution of information and communications technology platforms, organizations under the process of digital transformation constantly search for innovative concepts and creative applications. More importantly, ICIKS aims beyond the predominant role of technologies, seeking discussions on the essential interactions between individuals, who are the ultimate technology users.

Organizations must be aware of the importance of data, information, human behavior, and the knowledge owned by their members. Creating activities to enhance identification, preservation, and use of this knowledge is a powerful means to improve the level of an organization's performance. Thus, organizations must invest in human interactions. Since the COVID-19 crisis, more and more of these interactions are made through digital information systems which enhance knowledge sharing and cooperation culture. This will ensure better information and knowledge creation, transfer, and preservation taking into account the security of socio-technical systems.

We encouraged submission of research related to the design, implementation, and use of information and knowledge systems that take into consideration the cultural context of the organizations, and the viewpoints of their shareholders. Researchers and practitioners were invited to share recent advances in methodologies, models and tools concerning decision processes and information and knowledge systems that contribute to the success of digital transformation in organizations.

We received 32 papers from authors in a number of different countries. After an evaluation process by the Program Committee (PC) members, 13 research papers were selected: 10 full papers and 2 short papers. Each paper was double blind reviewed by at least three reviewers. These proceedings have been organized into three parts:

- Knowledge Systems and Decision Making;
- Machine Learning, Recommender Systems, and Knowledge Systems; and
- Security, Artificial Intelligence, and Information Systems.

We also had the pleasure of hosting four keynote speakers. Tung Bui from the University of Hawai'i (USA) discussed "Decision Paradigms and Support in the Age of New Normal"; Jean-Loup Richet from the Panthéon-Sorbonne University (France) dealt with "Security by Obscurity, a History of Secrecy"; Roland Inan, the Head of Information Systems (IS) and Information Technology (IT) Audit Assignments at FedEx (the Netherlands), dealt with "Data, knowledge management and cybersecurity in organizations"; and Leopold Larrios de Pina, the Head of Group Risk Management at Mazars (France) spoke about the "Digital era yes, but … data awareness first! It could be an introduction before taking risks".

ICIKS 2021 was organized in cooperation with the ACM (SIGAI and SIGMIS) and IEEE France section. It also received the support of the Amiens Business School, the University of Picardy Jules Verne, and the Paris Dauphine University (France), along with the University of Sfax (Tunisia). We are really grateful for their help and support.

We hope that this fifth edition of ICIKS was an important and privileged space for discussions and sharing experiences and innovation in the field of information and knowledge systems.

We would like to thank our Steering Committee, Program Committee and session chairs. Finally, we are grateful to the editors of the LNBIP series from Springer, Christine Reiss, Ralf Gerstner, and Alfred Hofmann, for their help in the preparation of these proceedings.

June 2021

Inès Saad
Camille Rosenthal-Sabroux
Faiez Gargouri
Pierre-Emmanuel Arduin

Organization

General Chair

Inès Saad   Amiens Business School and UPJV, France

Program Committee Chairs

Camille Rosenthal-Sabroux   University of Paris Dauphine – PSL, France
Faiez Gargouri   University of Sfax, Tunisia

Organizational Chair

Pierre-Emmanuel Arduin   University of Paris Dauphine – PSL, France

Steering Committee

Marie-Hélène Abel	UTC-Compiègne, France
Markus Bick	ESCP Business School, Germany
Fatima Dargam	SimTech Simulation Technology, Austria
John Edwards	Aston University, UK
Rim Faiz	University of Carthage, Tunisia
Michel Grundstein	University of Paris Dauphine – PSL, France
Gilles Kassel	University of Picardy Jules Verne, France
Shaofeng Liu	University of Plymouth, UK
Daniel O'Leary	University of Southern California, USA
Jean-Charles Pomerol	Sorbonne University, France

Program Committee

Marie-Hélène Abel	UTC-Compiègne, France
Laila Aligod	University of Sidi Mohamed Ben Abdellah, Morocco
Pierre-Emmanuel Arduin	University of Paris Dauphine – PSL, France
Sonia Ayachi	University of Sousse, Tunisia
Sana Ben Hamida	University of Paris Dauphine – PSL, France
Markus Bick	ESCP Europe, Germany
Wafa Bouaynaya	University of Picardy Jules Verne, France
Sarra Bouzayane	Audensiel, France
Salem Chakhar	University of Portsmouth, UK
Maria Manuela Cruz Cunha	Polytechnic Institute of Cávado, Portugal
Renata Paola Dameri	University of Genoa, Italy
Fatima Dargam	SimTech Simulation Technology, Austria
Claudia Diamantini	Marche Polytechnic University, Italy

John Edwards	Aston University, UK
Faiez Gargouri	University of Sfax, Tunisia
Mila Gasco	University at Albany, USA
Susan Gauch	University of Arkansas, USA
Sahar Ghrab	University of Sfax, Tunisia
Michel Grundstein	University of Paris Dauphine – PSL, France
Céline Joiron	University of Picardy Jules Verne, France
Gilles Kassel	University of Picardy Jules Verne, France
Ashraf Labib	University of Portsmouth, UK
Shaofeng Liu	University of Plymouth, UK
Florinda Matos	ISCTE – University Institute of Lisbon, Portugal
Nada Matta	UTT-Troyes, France
Nicholas Mavengere	Bournemouth University, UK
Brice Mayag	University of Paris Dauphine – PSL, France
Seif Mechti	University of Sfax, Tunisia
Davy Monticolo	University of Lorraine, France
Elsa Negre	University of Paris Dauphine – PSL, France
Káthia Marçal de Oliveira	University of Polytechnique Hauts-de-France, France
Camille Rosenthal-Sabroux	University of Paris Dauphine – PSL, France
Inès Saad	Amiens Business School and UPJV, France
María Isabel Sánchez Segura	Universidad Carlos III de Madrid, Spain
Muhammad Awais Shakir Goraya	University of Portsmouth, UK
Thierno Tounkara	Institut Mines-Telecom Business School, France
Eric Tsui	The Hong Kong Polytechnic University, China
Takahira Yamaguchi	Keio University, Japan
Sam Zaza	Middle Tennessee State University, USA
Saliha Ziam	University of du Québec, Canada

Additional Reviewers

Mousa Albashrawi
Abdullah Albizri
Yazan Alnsour
German Lenin Dugarte Peña
Ons Jallouli
Menatalla Kaoud

Myriam Merad
Simona Panaro
Halima Ramdani
Carlos Serrao
Dina Sidani
Sajid Siraj

Contents

Knowledge Systems and Decision Making

Expertise, Knowledge and Decision: Lessons from the Covid-19 Crisis in France

Jean-Charles Pomerol[✉]

Sorbonne Université and Agoranov, 96 bis Boulevard Raspail, 75006 Paris, France
`jean-charles.pomerol@sorbonne-universite.fr`

Abstract. This paper is devoted to the role and use of experts' knowledge for decision making. It is shown that experts' advices must be questioned because they disagree on many things in their field of expertise and are not immune of bias and conflict of interest. In few words, experts are generally not good decision makers and it is worst when they are assembled in a panel. The COVID-19 crisis provides many examples. Then we discuss how decision makers would use experts' knowledge.

Keywords: Decision making · Bias · Expert knowledge · Expert panel · Group thinking

1 Introduction

The competency of experts for making decision has already been discussed for a long time. More than thirty years ago Shanteau (1987, p. 290) states that the studies of experts' practice in various jobs, such that medical diagnosticians, stock brokers, clinical psychologists *"have consistently indicated that expert decision makers are inaccurate and unreliable"*: He adds that: *"The conclusions from earlier research is that experts are inadequate decision makers. This has been reinforced in more recent studies* (e.g., Chan 1982) *which have reported deficiencies in calibration (subjective-objective comparability) and coherence (internal consistency). Furthermore, expert decision makers are apparently unaware of these various shortcomings"*. In this paper we would like to review the reasons that can explain the discrepancy between expertise and decision. Furthermore, whereas experts possess an indisputable knowledge, the question of a wise use of this knowledge is addressed.

We will start by trying to understand what expertise is and who the experts are. We then examine the bias that undermines the reasoning of experts as well as panels of experts. We finish by giving some insight on the good use of expertise in decision making.

2 What Expertise is and Who is an Expert?

There are different types of expertise more or less encoded in experts' minds, depending of what is asked to an expert. To understand the difference, we have to think about the

© Springer Nature Switzerland AG 2021
I. Saad et al. (Eds.): ICIKS 2021, LNBIP 425, pp. 3–16, 2021.
https://doi.org/10.1007/978-3-030-85977-0_1

way the brain makes decisions. Basically human decision is recognition + reasoning (Pomerol 2012). Merging recognition and subsequent emotion with reasoning in the medio-prefrontal cortex is the characteristic of human decision (Damasio 1994). However, some decisions are "*recognition primed decisions*" (Klein 1993) or more precisely "*recognition primed action*", which means that in many circumstances an automatic action or reaction is triggered by the recognition of an image or a pattern. When you drive your car you make many instantaneous decisions without thinking about. This is the result of your experience: after a certain time of conscious training, the actions are encoded and no longer need reasoning. A skiing champion is undoubtedly an expert but his/her performance in a descent does not result from reasoning but from a very long training, it is the consequence of many internalized learnt moves. Any sport champion is representative of the first type of expertise.

At the opposite, some expertise rely almost totally on information and reasoning, for example lawyers' advices. Information and data bases of already judged cases, experience of similar cases and reasoning on the situation at hand is at the basis of the expertise of a lawyer. At this point, for this type of expertise, let us highlight a first difficulty. For a given case, you can always find a lawyer who gives you a good chance of winning your trial and a lawyer who says the same thing to the adverse party. The expertise would be unique there will be a much smaller number of trials! In insurance business it is commonplace: "*we have our experts and they have theirs!*" In any situation, medical, commercial, scientific, law, etc., it is always possible to find two experts with divergent advices. A specialist of judgment, Hammond (1987, p.2), quotes that "*experts disagree on virtually everything of importance or it seems so*". This is the case right now with the Covid-19 epidemiological expertise. We will see later how to manage this reality for decision making. At the beginning, in France, many experts, in particular the scientific council of the President, say that the mask was not useful for not infected persons, two months later the same advocate for mandatory mask for everybody. Certain epidemiologists on the basis of their models predict a second wave, whereas some others, on the basis of the H1N1 pandemic, claimed that the probability of a second wave was very low if not a pure phantasm.

Between expertise relying on reasoning and encoded expertise, there exists many possible mix. A good example is that of chess players. Chess game has been extensively studied by de Groot (e.g. de Groot and Gobet 1996). The expertise of masters relies on a number of recorded games, the recognition of topological structures on the chess board, and reasoning, especially pruning of the dead end moves. Chess game is a rare domain, except individual sport, in which expertise can be precisely and indisputably evaluated. It is unfortunately not the case in medicine, law, management, economics, policy and defense, six important domains in which the experts are often mobilized and more than often disagree…

The studies relative to chess masters have proved that what makes a difference between an expert and an inexperienced player is the structure of the data (de Groot and Gobet 1996). A master is able to record a situation by using some patterns structuring the pawn positions. It's probably the case for many other expertises: **an expert is somebody who has in mind a good structuring of the domain and is able to recognize patterns he/she has already met and recorded in his/her memory**.

The crux is that pattern recognition in decision making is a good and a bad thing! It's a good thing because it allows shortcomings in the reasoning and paves the way to rapid decision making, ultimately to *"recognition primed decision"* (Klein 1993). But the bad news is that recognition is conservative because it relies on situations that have been already encountered and recorded. This explains two reported drawbacks in expertise, the first one is that expert are very bad forecasters, see Makridakis (1990), Kahneman and Klein (2009), Kahneman (2011, ch. 20, 21 and 22). **Actually, the implicit assumption in recognition primed decision is that the environment has not changed since the recording of the case**. Predictions based on regression are often misleading in a nonstationary process or environment (Einhorn and Hogarth 1981; Kahneman 2011, ch. 21). Experts are good forecasters as soon as nothing changes in the future. One epidemiologist expert of the French scientific council for Covid-19, Professor Arnaud Fontanet, said in 2020 that no vaccine will be available before the end of 2021, it's just what could be inferred from past epedemics but with mRNA times changed! For the same reason panel of experts, for example in Delphi method, very rarely anticipate ruptures or inflexions. In a word, adopting Taleb's image (2007), black swan hunting is not a sport for experts.

The second consequence is that, as stated in Kahneman and Klein (2009), as soon as the environment is unstable or is very complex (*wicked*) with delayed feedback, recognition primed decision fails and consequently expertise must be questioned. For example, most of experts' advices in the Covid-19 epidemics rely on similarity with SRAS-Cov-1 and MERS-Cov, but it appears that Covid-19 is *wicked*, different from other corona virus, more contagious and to date we even don't know if the illness provides an enduring immunity. Never forget that most expertises implicitly assume a certain stability of a non-wicked environment, a *ceteris paribus* reasoning. The notion of stability of the environment is consistent with the results of Shanteau (1992, Table 2, p. 259) about task characteristics. He observes a good performance of experts in domains involving *"static objects or things"* (Shanteau 1992, p. 258). Snowden and Boone (2007) also think that experts are adapted to a complicated but ordered world.

It is difficult to characterize who an expert is. Shanteau (1987, Table 1) gave fourteen *"psychological characteristics of expert decision makers"*, among them let us retain the important ability to sort information in order to see what is relevant or not and the capacity of structuring complicated or complex situations. The fact that a part of the expert's knowledge is encoded in his/her brain has many consequences which are mentioned by Shanteau (1987, 1988): experts need less time and effort to make a decision than novices. They exhibit a greater automaticity in the decision process than novices but they are often unable to explain precisely their decision because an encoded reaction is difficult to decipher. Finally, it has been noted that experts have a greater stress tolerance than inexperienced people probably because of *"a greater automaticity in the cognitive processes"* (Shanteau 1987, p.296). Other psychological traits will be seen in the next section.

3 Bias in Expertise

There are many biases in decision making that are known since the fundamental works of Kahneman and Tversky; see Kahneman (2011) for an extensive survey and Pomerol

(2012) for the consequences in decision making and action. In this section, I limit myself to the biases that heavily impact expertise.

3.1 Overconfidence

Overconfidence in its own expertise is not reserved to experts, numerous experiences have proved that everybody might be subject to be overconfident in his/her judgment or evaluation, see Kahneman and Lovallo (1993), Kahneman et al. (1982, ch. 20, 21, 22), Kahneman and Tversky (2000, ch. 22, 23), Michailova and Albrechts (2010 and the reference therein) and Kahneman (2011, ch. 20). Overconfidence has some link with the well-known illusion of control (Kahneman et al. 1982, ch. 16; McKenna 1993) which wrongly induces people to think that they control events. What is striking is that experts suffer the same overconfidence bias as most people, moreover one can say that generally, more expert you are, more you are mistaken with confidence, because overconfidence as well as the illusion of control increase with the difficulty of the task (Michailova and Albrechts 2010). However, experts are different from ordinary people who are more confident in their judgment when they are ignorant or beginners in a domain, than when they are really knowledgeable which is known as the Dunning-Kruger effect (Dunning 2011), on the contrary experts' overconfidence increases with their knowledge (e.g. Michailova and Albrechts 2010). The problem is that, still different from ordinary people, experts are not prone to the second Dunning-Kruger effect *i.e.* that *"top performers tends to underestimate their performance"* (Dunning 2011, p. 270 *et seq.*). *"The experts have, after all, invested in building their knowledge and they are unlikely to tolerate controversial ideas"* (Snowden and Boone 2007). Overconfidence in his/her own judgment is difficult to overcome because it's almost impossible to learn from outcomes (Einhorn and Hogarth 1978) as it is well-known in decision theory.

The overconfidence of experts may have dramatic consequences, some of them are well documented, see *e.g.* Janis and Mann (1977) and Morel (2002, 2012). In many cases the expertise is shared by several experts, as we will see in the next section, and the responsibility of the disaster is shared with the hierarchy. But accidents resulting from one unique expertise are also documented; it's the case in mountain sports. McCammon (2002) has observed that skier groups are more often victims of avalanches when they are led by experienced leaders. More generally experts in one activity tends to take more risks than novices, it's obvious in dangerous sports but is also the case in many professional settings. As observed by McCammon and before by many authors (e.g. Heinrich 1931; Reason 1997) it's also a consequence of the familiarity with the context, be it dangerous.

Illusion of control and overconfidence are, like Charybdis and Scylla the twin reefs that mostly jeopardize expert advices and their decisions. But there are other biases that may influence experts.

3.2 Other Biases

Anchoring and framing effects are two biases perfectly documented by Kahneman and Tversky (2000), it is not impossible that some experts may be exposed to them, but it is probably relatively rare because experts have a good knowledge of their domain which prevents them to be lure by a deceitful wrapping of the facts.

It's probably different for two biases linked to memory because of the large use of recognition in expertise. The first one is the bias of representativeness and the second is availability (Kahneman et al. 1982). The judgment of experts mainly relying on their experience, it is plausible that the situations remaining salient in their memory are more easily recorded and used for future expertise. Nobody is immune to the *"peak and end rule"* (Kahneman 2011). Thus, the availability effect incites to conservatism and prevent from exploring new paths. But representativeness is even more dangerous for expertise because it impacts the probabilities. What is difficult to be represented in your mind will be endowed with smaller probabilities than real. As said by Thomas Schelling who was a famous expert in defense policy (1962, p. vii) *"There is a tendency in our planning to confuse the unfamiliar with the improbable. The contingency we have not considered looks strange. What looks strange is thought improbable: what is improbable need to be considered seriously"*. His recommendation remains of the first importance for experts. There are numerous examples in the literature of small probabilities, of the order of magnitude of 10^{-2} to 10^{-3}, which have been neglected by experts, *e.g.* the exceptionally low temperature in the Challenger accident (Morel 2002), or the centennial wave in Fukushima, without evoking Pearl Harbor (Schelling, 1962; Janis and Mann 1977) and the Bay of Pigs (Janis 1982) as well as many other historical examples.

Overconfidence is probably the most frequent experts' trait, which raises the question of learning and of the possibility for an expert to be conscious of his/her limits. Unfortunately, it is difficult to learn because the result is not an adequate indicator of the quality of the decision. The question of how much experts learn from feedback is documented and discussed at large in Einhorn and Hogarth (1978, 1981), Kleinmuntz (1985, 1993), Axelrod and Cohen (2000). They point out that it is very difficult for an expert to assess his/her judgmental accuracy in most real settings. The value of the outcome which depends on the events is not an indication of the quality of his/her decision, perhaps if the decision is more or less repetitive and that the outcomes remain good, one can begin to believe in the quality of the decision maker. Definitely, the result is not a good indicator of the quality of the decision and a good result occurring by chance is very often misleading (*e.g.* Axelrod and Cohen 2000). The difficulty of learning from outcomes, with memory biases paves the way to the illusion of validity (Kahneman 2011, Ch. 20). The structure of the task also makes it difficult to learn because negative judgments are often not available to study. For example, a recruitment expert never knows whether the rejected candidates would have been better than the retained ones. This is one component of the illusion of validity (Kahneman and Tversky 1973; Einhorn and Hogarth 1978). Already, according to Cicero, Diagoras answered to somebody who argues that thousands of *ex-voto* in temples prove the existence of divine interventions: *"but all who sunk haven't had the opportunity of a votive writing!"* (Mentioned by Taleb 2007). Neglecting the size of the sampling or of the base rate is a very common mistake (Einhorn and Hogarth 1978; Kahneman 2011, ch.16) entailing that it is even more difficult to learn probabilities. Einhorn and Hovarth (1978) stated that: *"The difficulty of learning from experience has been traced to three main factors: (a) lack of search for and use of disconfirming evidence, (b) lack of awareness of environmental effects on outcomes, (c) the use of unaided memory for coding, storing and retrieving outcome information.*" The first and the third reasons are related to the well-documented confirmatory bias inducing that what confirms a judgment or a decision is preferentially stored in memory feeding the illusion of validity.

Finally let us say some words about conflicts of interest. Let us just remind that there were, for many years, experts who claim in well documented reports that asbestos and tobacco were not dangerous. Experts paid by oil companies still deny climate change. Those paid by pharmaceutical companies defend the company's drugs. These are obvious examples of conflict of interest in expertise. But I would like to point out a less obvious phenomenon which is clearly visible in the Covid-19 epidemics. In 1987, an expert in judgment Kenneth Hammond already wrote (1987, p.2):" *Our confidence in expertise is also diminished by the obvious and all too willing exploitation of experts by government and large corporations, as well as their opponents, circumstances in which experts appear to be no more than hired guns*".

In centralized countries like Russia, China, France and many others the researchers' resources depending almost exclusively on the state and its agencies, it is then difficult to believe in an independent expertise. Covid-19 crisis brings us many examples. In France the scientific council of the President says, in March, that it was necessary to ascertain the availability of surgical masks for the population without any recommendation to wear them[1]. The reason appears later because France has no stock of masks, and the same experts recommend in April that mask wearing becomes compulsory in public transportation (20/4/2020). A recent study by professor Yuen Kwok-Yung[2] of university of Hong Kong proved with no doubt that masks were useful for prevention!

The head of this scientific council, Jean-François Delfraissy, a top epidemiologist claims, on the basis of council "reflections", in April that we ought not unconfined the elders for months while in May it was no longer necessary because politically impracticable and probably unconstitutional. In France, only one virologist expert, Didier Raoult, advocates for largely testing people in order to limit the propagation of the virus. Almost all the French official experts were for a limitation of the test to patients having manifest symptoms of Covid-19. The only reason was that the French government has almost no test and very low testing capabilities in public hospitals and does not want to involve private laboratories in a testing campaign. I guess that we could multiply the examples in many countries. It is very questionable that experts integrate in their advice some considerations outside the scientific field, for example economic or political. Added to the fact that the experts' advices are diverse, sometimes divergent because the knowledge is in construction, this discredits expertise and even brings science into disrepute.

As we just observed with the scientific council, the fact that many experts are assembled in a panel does not result in a better advice, it's often worse as we will explain in the next section.

---

[1] In the same advice, see https://solidarites-sante.gouv.fr/actualites/presse/dossier-de-presse:article/covid-19-conseil-scientifique-covid-19, the committee explicitly says that it takes into account some political and economic principles.

[2] https://www.i24news.tv/fr/actu/international/asie-pacifique/1589738086-hong-kong-corona virus-des-tests-sur-les-hamsters-prouvent-l-efficacite-des-masques-experts.

4 Expert Panels

One can naively think that while experts' advices may be divergent, it is wise to make decision or ask advice to a panel of experts. This is unfortunately not the case; it is often worse. The mathematician geometer André Lichnerowicz used to say that: *"the cleverness of a panel is vectorial, each vector is large but the sum can be null"*. He experiences this adage when working on school programs of "new" mathematics in the seventies, many experts intervene but, at the end, the programs were practically unfeasible. One of the reasons is, that each expert tries to promote what he/she thinks is absolutely essential, but what is essential differs from an expert to another, resulting in an enormous corpus. For example, the recommendation booklet[3] written by the government agency AFNOR to make at home a mask against Covid-19 is 36 pages long! The booklet for reopening the classrooms is about ninety pages long. A second reason is that, in a panel of experts, everybody watches and is jealous of the others, each of them wants to prove that he/she is smart enough to add something; it's a question of prestige. The ego battle can also result into "analysis paralysis" (Snowden and Boone 2007).

Studies in group dynamics go back to Kurt Lewin (1947). Group thinking was then extensively studied; see for example Janis (1982) for a synthesis. There are several biases in group decision. Among them there are the well-known experiences of Asch (1952). The authority of a group leader induces a change of the advice of a member of the group to conform to the majority even when the majority is wrong, while he/she is obviously right: conforming to dominant advice is one of the pitfalls in group deliberation. Kurt Lewin (1947) shows the reinforcement of opinions into groups and the commitment to the majority. Expert groups are not immune to this phenomenon. For example, in the Challenger shuttle accident it seems that few persons draw the attention to the possible defect of booster joins at low temperature, but they were ignored by the group and the hierarchy (Morel 2002). In the Bay of Pigs decision all the CIA experts agree on the feasibility of the operation: obvious group thinking. A small probability event is almost surely discarded by a group of experts. Between the two world wars, the defense experts in France advocate for the fortification of the east frontier (*ligne Maginot*) whereas the German army violated the Belgian neutrality in 1914 and the probability that they do it again was not very small! Expert group blindness. Moscovici (1976) speaks of group polarization. **Commitment to the group and polarization are two group effects that are present in expert groups.** In a group of defense experts, no one wants to be considered timorous. Once more it is difficult to oppose to a group (Lewin 1947) and the polarization effect leads to a reinforcement of the group belief. A similar effect entails group extremeness (Moscovici and Zavalloni 1969; Sunstein 2000) which occurs for example in political parties. As an example of polarization, the scientific council mainly composed of epidemiologists and virologists gave a unanimous advice not to re-open the schools before September (20/4/2020), whereas the council of the child psychologist society unanimously says that it was of the most importance to re-open now! Depending on the decision you have in mind, you can find the expert advice that fits well!

Another effect is the risky shift (Moscovici and Zavalloni 1969) which leads a group of experts to make riskier decision than each expert separately. It's obvious in defense

---

[3] AFNOR SPEC-576–001.

and is reinforced by another phenomenon that is neglecting independent probability multiplication. Consider an expertise on a composite task, such that rescue of American hostages in the Iranian embassy (operation *Eagle Claw*), each subtask having a reasonable chance of success may result in a weak probability of success for the whole operation. Actually, in the case of the American hostages of the Iranian embassy, the study by Rosenzweig (1993) shows that there were five subtasks with probabilities of success between 85% and 55% resulting in a 11% chance of success. Multiplying independent probabilities decreases rapidly the chance of success. This phenomenon is often minimized in group reflection, each expert being specialized on a subtask; he/she is confident in his/her expertise and does not envision the chaining. The same thing occurred with mathematical teaching curricula or the mask fabrication instructions, each expert has a specialty and nobody says that the total is nonsensical for students or seamstress.

The opposite to the risky shift can also be observed in expert groups, the so-called "*precautionary principle*" entails each expert to be particularly cautious, even timorous, resulting in a total helplessness. In the Covid-19 crisis this is the case, *e.g.* once more consider the scientific council of President Macron, and its suggestion of not re-opening the schools before September 2020 (cf. supra). Note that nobody knows, even these experts, whether Covid-19 will be disappeared in September or will recur. The same principle recently leads World Health Organization (WHO) to suspend clinical trials on hydroxichloroquine (May 25th 2020) on the basis of a discussible *Lancet* publication. WHO is immediately followed by the French scientific council. But is there only one clinical trial possible respecting the "precautionary principle"? Finally, a lot of scientists question the methodology of the study and three of the four authors of the study retracted resulting in the withdrawing of the paper. Another illustration of the "precautionary principle" is given by the dramatic difference between French and American policy about vaccines. Relying on the belief of some experts that no vaccine is rapidly possible (see supra), French government made no investment and no command in startups working on vaccines, while USA and UK invested more than 9 billion in biomedical startups.

Whereas it is difficult for an expert to learn from his/her past decisions or recommendations, it is even more difficult for a group, supposing that organizational learning is possible which is discussed for a long time (*e.g.* in Simon 1997, ch. VIII; March and Olsen 1975). Einhorn and Hogarth (1981) ask: "*what, if anything did the US learn from the Viet Nam war?*" I should be tempted to answer "nothing" at the view of the posterior events in the Middle East. But, some persons learnt, for example, in this case, Robert McNamara, see his book (1995). After the department of defense, he headed the Ford division in the Ford company where he stopped the sunk investment in the Edsel model, proving that he has understood the "*too invested to give up*" trap (Zaleznik 2005). **Experts can sometimes learn, panel of experts never**!

One can say the same thing of the Bay of Pigs disaster, it is questionable to say that the CIA has learnt but undoubtedly John F. Kennedy did. He understood that experts in group can be wrong and that it was necessary to develop a culture of discussion and study of the non-conventional opinions, the dominant advice must not be the only one to be documented. The lesson was very useful in the Cuba missile crisis (Allison 1971; Janis and Mann 1977; Janis 1982). The second lesson for John F. Kennedy was that the progressivity and reversibility of the decision is worthwhile and generally better than one shot risky bet. The "two irons in the fire" policy and irremediable decision postponing as long as possible are two very important figures in practical decision making under uncertainty (Pomerol 2001).

The value of a group relies in the weighing of pros and cons, provided that the con arguments are not stifled. The existence of adversary opinions does not suffice *per se* to generate a good deliberation. A group is not naturally open-minded to challenging information; the cohesion shuts the door to contradictors (Janis and Mann 1977, about Pearl Harbour). Diversity is not sufficient (Lev-On and Manin 2009), the debate must be contradictory. It is as much as necessary that the confirmatory bias entails that everybody, including experts, tends to retain only the arguments and facts that confirm their opinion. The same phenomenon occurs on Internet, people are attracted by like-minded sites (Lev-On and Manin 2009; Bronner 2016). Hence, deliberation and contradictory discussion must be organized, which means that devil's advocates must be heard, understood and weighed. Contradictory discussion is the price for security and catastrophic decisions avoidance (Morel 2002, 2012).

5   Why the Knowledge of Experts is Necessary but not Sufficient for Decision Making

"No lesson seems to be so deeply inculcated by experience of life as that you never should trust experts"

Lord Salisbury 1877[4].

We have seen that expert reasoning is not free of biases and often tainted by conflict of interest. It's a first reason to be reluctant to follow expert recommendations. But let us turn back the question: why expert advices are nevertheless important? As we have said an expert's advice relies on the best knowledge in a narrow domain, more precisely on what already happened in this domain adapted to what can be anticipated. In an expertise, what is important is not the final advice but the arguments. **A good decision maker must listen to the arguments of the experts, understand them and follow his/her rationales up to their advice**. And more important the decision maker must question the assumptions of the expert on the environment. It is very important to question the stability of the environment, the speed of its change and the availability of not too delayed feedback, in a word whether the environment is complex or complicated in the terminology of Snowden and Boone (2007). It is also most important to decipher the anticipations of the expert, in decision theory words: what are his/her subjective probabilities. It's often difficult because these probabilities are generally not made explicit. For example, if an expert recommends to build a new factory, it is important to know what are his/her assumptions on the price of energy within 10 years, an oil barrel below 50 $ or higher that 80 $ with attached probabilities. **Nobody knows but the expert has necessarily guessed it.** Experts must unveil their assumptions and anticipations.

One thing to consider is that, as said by Shanteau (1988, p. 204), experts often appear *"as cognitively limited decision makers"*. They are limited to their domain. But in many

---

[4] Quoted by James Shanteau (1992).

operational decisions, in business and public affairs, the domain is not so narrowly delimited. The decision maker has to consider a large landscape of facts and a large span of time. This holistic view is not usual for experts. Moreover, the criteria for decision making are multiple. For example, in the Covid-19 crisis, it is obvious that one criterion is the healthiness of the population, but another one is the economic consequences of confining people. In practical decision making, despite politicians' affirmations, there never exists an optimal decision, the components of the choice, or criteria, being contradictory (Pomerol and Barba-Romero 2000). In some countries like USA and Brasil, the number of deaths has not been weighted as much as economic depression risk and other electoral considerations. More generally in any practical decision making the conflict between long-run and short-run objectives is inevitable (Pomerol 2012), but it's not the role of experts to compromise. As recalled by Philipps-Wren et al. (2020), following Simon (1997), one does not confuse facts and values: facts are the realm of experts, values of decision makers.

Invoking experts' advice in a muticriterion setting is nothing else than trying to hide his/her decision behind alleged scientific rationales, or worse, not being able to assume any decision. Once more, the good use of experts is not to let them debate in a council who provides advices flawed by group thinking biases and conflict of interest with the power, but to heard divergent expertise and organize a contradictory debate (public or confidential) before decision making. Acting otherwise is political communication or pedagogical sincere desire to inform people of the complexity of the things. Bill Below OECD Directorate for Public Governance and Territorial Development quoted the alleged citation of the satirist H.L. Mencken: "For every complex problem there is an answer that is clear, simple and wrong"[5]. In this time of Covid-19, it is useful to remember it, populists and self-proclaimed experts could adopt Mencken adage.

What appears when listening to many experts in this Covid-19 crisis is that they rarely say "we don't know". We don't know whether the epidemics will recur in winter, we don't know why many people have no symptom, we don't know why the virus impacts preferentially elder people and men, we don't' know whether 20% of the population being infected (actual figures) is sufficient to stop the diffusion of the virus, whereas some epidemiologists say that 70% is necessary, and so on… There are so many questions. Nevertheless, each expert perseverates in his/her conviction. Perseveration bias is well-documented in certain contexts (aviation, alpinism), see Dehais et al. (2010) and Morel (2012). Perseveration is reinforced in stressful and unexpected situations (Dehais et al. 2020), this seems to be the case now. Perseveration about hydroxichloroquine effects is the rule for disclaimers and defenders of the miraculous pill. This is another reason why experts are bad decision makers; a good decision maker must know how to lose (Pomerol 2012). An expert is always absolutely confident in his/her knowledge.

Beyond inherently involving multiple criteria, **decision making for action is definitively not only judgment** as pertinently observed by Einhorn and Hogarth (1981). Action implies commitment, personal involvement, sense giving by a hearty wording and good arguments (Pomerol 2012). Einhorn and Hogarth (1981) say that: "Unlike judgments, actions are intimately tied to notions of regret and responsibility". A good decision

---

[5] http://oecdinsights.org/2017/02/07/out-of-complexity-a-third-way/ many thanks to Jean-Claude Andé who draws my attention to this paper.

maker assumes the responsibility of his/her decision, firstly the responsibility of helping and protecting his/her followers; also the responsibility with regards to observers and opinion. On the contrary an expert engages his/her responsibility about the reliability of the diagnosis and the quality of his/her knowledge but does not assume the consequences of the advice that is possibly drawn by somebody from his/her knowledge. It makes a big difference with decision making. It has been obvious during the Covid-19 crisis, many "experts" advice of confining or not people, using hydroxichloroquine or not, opening the schools or not, and so on, but let the responsibility of the decisions to the government and none of them insisted to take the blame.

Responsibility also implies that the decision maker weighs not only the pros and cons but also the risk. There is no action absolutely without risk. Pomerol (2012) estimated that people usually accept the risk of unfavorable events with probabilities between 10^{-1} and 10^{-2} when the issue at stake, in case of bad luck, is slightly unfavorable. When the result is unfavorable or worst involves a "chance" of ruin (or death) almost nobody bets with a probability larger than 10^{-3}. This magnitude is accepted only in risky sports thanks to the illusion of control. The chance of a car accident is between 10^{-4} and 10^{-5} in many countries, which deters nobody to get a car, also thanks to the illusion of control or "*it will never happen to me*" bias (Kahneman et al. 1982, Ch. 33).

When making a decision, it's necessary to consider moderately unfavorable events with probabilities larger than 10^{-3} but from 10^{-4} and below, even in ordinary life, action should not be impeded. We retrieve these magnitude orders in the Covid-19 crisis. Actually, it seems that the probability of being infected to death with Covid-19 is about 2 10^{-3} for youngers, it's within the limit of an acceptable risk with the illusion of control; this probably explains why the shielding measures are poorly accepted by youngers. However, the risk is not uniformly spread, it varies from 3% to almost 20% for elderly people from sixty to ninety and more. This is the reason why some experts consistently propose that elder people be confined for months, which was considered as politically unacceptable. Once more, risk management proves that expertise and decision making are different jobs. For an epidemiologist expert like Jean-François Delfraissy, seeking for the minimal risk for elder persons enforces to lock them up. But nobody wants to endorse such a decision because even older people vote and do not want to live at zero risk!

6 Conclusion

Hamond (1987) begins by: "*One of the marks of the 20th century must surely be the rise of the importance of expert.*"; We are now surrounded by an army of experts[6]. This is exactly the case in the Covid-19 crisis, while the knowledge about the virus is sparse! This epidemics episode perfectly illustrates the instinctive feeling of many decision makers and psychologists who studied expertise and judgment like Janis, Shanteau, Hohgarth, Hammond, Tversky, Kahneman and others, that expert judgments are questionable and

---

[6] The UK politician Michael Grove said about the Brexit and expert economists: "*the people of this country have had enough of experts from organisations with acronym saying that they know what is the best and getting it consistently wrong*". https://www.newsweek.com/michael-grove-sky-news-brexit-economics-imf-466365.

that policy makers ought seriously to try to make their own opinion rather than blindly trusting experts, even assembled in a panel. In this paper we have tried to explain why this feeling is explainable by the usual mind biases, firstly overconfidence, then perseveration, finally reinforced in a panel by group thinking.

The qualities of a good decision maker (see Pomerol 2012 for a review) are not the same than those of an expert. What we expect of a good expertise is precise, up to date, documented knowledge and valuable arguments. Apart of human qualities like empathy, sociological instinct, loyalty, in decision making, we must value the capacity to anticipate and understand other reactions, to assess the risks, to drive teams, to react quickly and wisely.

However, experts and their knowledge are necessary for decision makers. There is a good use of expertise. **Experts are not the problem but the fact of relying on a unique expertise**[7]. Therefore, the decision makers must listen to experts' advices, knowing that they are diverse, sometimes contradictory. They have to weigh the pros and cons arguments keeping in mind that the multiple criteria of the choice are not of the experts' relevance. They also have to be able to distinguish between fake and auto-proclaimed experts in search for visibility, and established science[8]. Knowing who is who and where to find the information is a competency of managers. These are some of the reasons why decision making is not an expert's job. Also, in order to reveal the hidden assumptions of the experts, especially as regards future, probabilities and environmental changes, it is useful to organize contradictory debates when possible, and otherwise to systematically listen to divergent expertise and value the so called devil's advocate (Morel 2012). As recalled by André and Masse (2003) expertise is like Aesop's language the best thing when it is used wisely, understanding the hidden assumptions, the environmental uncertainties and changes and the knowledge limitations, and the worst when it is accepted without examination and used to infantilize the public[9] or divesting his/her responsibilities. We are obliged to observe that during the Covid-19 crisis many governments have mobilized experts, or expert panels like in France, to justify their decisions and not incurring the blame for restricting the liberties or for an upsurge of the epidemics, thus increasing cacophony and discrediting experts' advices.

References

Allison, G.: The Essence of Decision, Explaining the Cuban Missile Crisis, Little Brown, Boston (1971)

André, J.-C., Masse, R.: L'expertise La Science et l'Incertitude : l'Expertise Scientifique ou la Langue d'Esope? Environnement Risques & Santé 1(5), 299–306 (2003)

---

[7] Niall Ferguson, interview to the French weekly magazine Le Point: *"La nouvelle guerre froide a du bon"* 06/05/2020.

[8] In a well-documented paper Pascal Marichalar retrace the story of the epidemic from December 2019 to now. He proves that the *Science journal* gave an excellent real time follow up of the epidemic and announced the pandemics on the 25th February. Provided that countries had taken containment measures at that time, the casualties and the damages would have been much lesser.

[9] See Bernadette Bensaude-Vincent, https://theconversation.com/penser-lapres-sciences-pouvoir-et-opinions-dans-lapres-covid-19-137272.

Asch, S.: Social Psychology. Prentice Hall, Englewood Cliffs (1952)

Axelrod, R., Cohen, M.D.: Harnessing Complexity. Basic Books, New York (2000)

Bronner, G.: Belief and Misbelief, Asymmetry on the Internet. ISTE-Wiley, London (2016)

Chan, S.: Expert judgments made under uncertainty: some evidence and suggestions. Soc. Sci. Q. **63**, 428–444 (1982)

Damasio, A.R.: Descartes' Error. Putman's Sons, New York (1994)

Dehais, F., Lafont, A., Roy, R., Fairclough, S.: A neuroergonomics approach to mental workload, engagement and human performance. Front. Neurosci. **14**, Article 268 (2020)

Dehais, F., Tessier, C., Christophe, L., Reuzeau, F.: The perseveration Syndrome in the Pilots' Activity: guidelines and cognitive countermeasures. In: Palanque, P., Vanderdonckt, J., Winkler, M. (eds.) Human Error, Safety and Systems Development, pp. 68–80. Springer, Berlin (2010). https://doi.org/10.1007/978-3-642-11750-3_6

Dunning, D.: The Dunning-Kruger effect: on being ignorant of one's own ignorance. In: Zanna, M., Olson, J. (eds.) Advances in Experimental Social Psychology 11, pp. 247–296. Elsevier (2011)

de Groot, A.D., Gobet, F.: Perception and Memory in Chess: Heuristics of the Professional Eye. Assen, Van Gorcum (1996)

Einhorn, H.J., Hogarth, R.M.: Confidence in judgment: persistence of illusion of validity. Psychol. Rev. **85**, 395–416 (1978)

Einhorn, H.J., Hogarth, R.M.: Behavioral decision theory: processes of judgment and choice. Annu. Rev. Psychol. **32**, 53–88 (1981)

Hammond, K.R.: Toward a unified approach to the study of expert judgment. In: Mumpower, J.L., Phillips, L.D., Renn, O., Uppuluri, V.R.R. (eds.) Expert Judgment and Expert Systems, pp. 1–16. Springer, Berlin (1987). https://doi.org/10.1007/978-3-642-86679-1_1

Heinrich, H.W.: Industrial Accident Prevention: A Scientific Approach. McGraw-Hill, New York (1931)

Janis, I.L., Mann, L.: Decision Making, A Psychological Analysis of Conflict, Choice and Commitment. The Free Press, New York (1977)

Janis, I.L.: Groupthink: Psychological Studies of Policies Decisions and Fiascoes. Houghton Mifflin, Boston (1982)

Kahneman, D.: Thinking Fast and Slow. Allen Lane, London (2011)

Kahneman, D., Klein, G.: Conditions for intuitive expertise, a failure to disagree. Am. Psychol. **64**(6), 515–526 (2009)

Kahneman, D., Lovallo, D.: Timid choices and bold forecast: a cognitive perspective on risk taking. Manag. Sci. **39**, 17–31 (1993)

Kahneman, D., Slovic, P., Tversky, A.: Judgment Under Uncertainty: Heuristics and Biases. Cambridge University Press, Cambridge (1982)

Kahneman, D., Tversky, A.: On the psychology of prediction. Psychol. Rev. **80**, 231–273 (1973)

Kahneman, D., Tversky, A.: Choices, Values and Frames. Cambridge University Press, Cambridge (2000)

Klein, G.A.: A recognition primed decision (RDP) model of rapid decision making. In: Klein, G.A., Orasanu, J., Calderwood, R., Zsambok, C.E. (eds.) Decision Making in Action, Models and Methods, Ablex, Northwood, pp. 138–147 (1993)

Kleinmuntz, D.N.: Cognitive heuristics and feedback in a dynamic decision environment. Manag. Sci. **31**, 680–702 (1985)

Kleinmuntz, D.N.: Information processing and misperceptions of the implication of feedback in dynamic decision making. Syst. Dyn. Rev. **9**, 223–237 (1993)

Lev-On, A., Manin, B.: Happy accidents: deliberation and online exposure to opposite views. In: Davies, T., Gangadharam, S.P. (eds.) Online Deliberation: Design, Research and Practice, pp. 105–122. CSLI Publications, University of Chicago Press (2009)

Lewin, K.: Frontiers in group dynamics, concept, method and reality in social science: social equilibria and social change. Hum. Relat. **1**(1), 5–41 (1947)

Makridakis, S.G.: Forecasting, Planning and Strategy for the 21th Century. The Free Press, New York (1990)

March, J.G., Olsen, J.P.: The uncertainty of the past: organizational learning under ambiguity. Eur. J. Polit. Res. **3**, 147–171 (1975)

McCammon, I.: Evidence of heuristics traps in recreational avalanche accidents. Presented at the International Snow Science Workshop, Penticton, British Columbia, 30 September–4 October (2002)

McKenna, F.P.: It won't happen to me: unrealistic optimism or illusion of control. Br. J. Psychol. **84**, 39–50 (1993)

McNamara, R.S.: In Retrospect: the Tragedy and Lessons of Vietnam. Random House, New York (1995)

Michailova, J., Albrechts, C.: Development of the Overconfidence Measurement Instrument for the Economic Experiment. MPRA paper (2010). http://mpra.ub.uni-muenchen.de/26384/

Morel, C.: Les décisions absurdes. Galimard, Paris (2002)

Morel, C.: Les décisions absurdes II, comment les éviter. Galimard, Paris (2012)

Moscovici, S.: Social Influence and Social Change. Academic Press, New York (1976)

Moscovici, S., Zavalloni, M.: The group as polarizer of attitudes. J. Pers. Soc. Psychol. **12**, 125–135 (1969)

Phillips-Wren, G., Pomerol, J.-Ch., Neville, K., Adam, F.: Supporting decision making during a pandemics: influence of stress, analytics, experts, and decision aids. In: Liebowitz, J. (ed.) The Business of Pandemics: The Covid-19 Story. Taylor and Francis, Boca Raton (2020)

Pomerol, J.-C., Barba-Romero, S.: Multicriterion Decision Making in Business. Kluwer, New York (2000)

Pomerol, J.-C.: Scenario development and practical decision making under uncertainty. Decis. Support Syst. **31**, 187–204 (2001)

Pomerol, J.-C.: Decision Making and Action. ISTE-Wiley, London (2012)

Reason, J.: Managing the Risks of Organizational Accidents. Ashgate, Aldershot (1997)

Rosenzweig, P.M.: Judgment in Organizational Decision Making. The Iranian Hostage Rescue Mission', Division of Research. Harvard Business School (1993)

Schelling, T.C.: Pearl Harbour – Warning and Decsion. Stanford University Press, Stanford (1962)

Shanteau, J.: Psychological characteristics of expert decision makers. In: Mumpower, J.L., Phillips, L.D., Renn, O., Uppuluri, V.R.R. (eds.) Expert Judgment and Expert Systems, pp. 289–304. Springer, Berlin (1987). https://doi.org/10.1007/978-3-642-86679-1_16

Shanteau, J.: Psychological characteristics and strategies of expert decision makers. Acta Physiol. **68**, 203–215 (1988)

Shanteau, J.: Competence in experts: the role of task characteristics. Organ. Behav. Hum. Decis. Process. **53**, 252–266 (1992)

Simon, H.A.: Administrative Behavior, 4th edn. The Free Press, New York (1997)

Snowden, D.J., Boone, M.E.: A leader's framework for decision making. Harv. Bus. Rev. **85**(11), 68–76 (2007)

Sunstein, C.R.: Deliberative troubles? Why groups go to extremes? Yale Law J. **110**, 71–119 (2000)

Taleb, N.N.: The Black Swan, the Impact of Highly Improbable. Penguin Books, London (2007)

Zaleznik, A.: The education of Robert S. Mcnamara, secretary of defense 1961–1968. Rev. Fr. Gest. **159**, 45–70 (2005)

E-learning: Factors Affecting Students Online Learning During COVID-19 Quarantine in a Developing Country

Nada Mallah Boustani[(✉)] [ID] and May Merhej Sayegh[(✉)] [ID]

Faculty of Business and Management, Saint-Joseph University, Beirut, Lebanon
{nada.mallahboustany,may.merhejsayegh}@usj.edu.lb

Abstract. In response to the impending spread of COVID-19, universities world-wide switched to digital learning. They switched to online platforms using technology in their learning processes. In this context, an online survey study was conducted in Saint Joseph University, Faculty of Business in which the authors collected 458 responses from students during the spring-summer 2020 semester. The quantitative study focused on (1) the students' access to technology, (2) the facilitating conditions offered by the university during the semester (3) the learning value acquired from the online teaching and (4) the demographics characteristics as gender and level of education. The researchers tested the relationship between those variables and the intention to use e-learning and to continue using it after the pandemic crisis. In this paper, the authors present and discuss their study's results, where no gender differences were noted but a disparity in the level of education led to acceptance or rejection of future use of e-learning; moreover, the access to technology raised a major problem specially in the case of a developing country.

Keywords: COVID-19 · eLearning · Access to technology · Learning value

1 Introduction

Digital technologies have provided support in diverse sectors and areas in the COVID-19 outbreak in general and in education specifically [1]. Higher education was affected radically in this period. There is an increasing use of e-learning in the educational institutional and universities with the spread of the pandemic crisis all over the world [2]. With the widespread use of e-learning platforms starting 2020, many researchers tried to understand the factors impacting the acceptance and the intention to continue using e-learning even after the pandemic crisis in the future. In order to better understand the factors influencing students in educational institutions, this research considered the possible influence of a range of factors related to the learner such as demographics factors (age, gender and level of education) beside three more variables: e-learning value, facilitating conditions offered by universities and access to technology available in the Lebanese context.

It is often claimed that younger generation will be more accepting the use of Technology in learning [3]. University students are considered to switch easily to digital

© Springer Nature Switzerland AG 2021
I. Saad et al. (Eds.): ICIKS 2021, LNBIP 425, pp. 17–28, 2021.
https://doi.org/10.1007/978-3-030-85977-0_2

technologies and the virtual environment and should not present any operational problem [4]. Researchers found a new field of interest with the unprecedented situation generated by the Coronavirus pandemic, One of their studies was the impact of the e-learning adoption on Business school students and what are the factors that can affect the intention to use e-learning in the future. Two main questions will be explored in our study: What is overall E-Learning Acceptance among the students, what is the learning value of it? What is the role of demographics (gender, educational level) in E-Learning intention to use?

2   Theoretical Background

E-learning is the adoption and use of computers, network and communication technologies with graphics, videos and audio [5]. The COVID-19 crisis has modified people's habits and had obliged students to change their method of learning. Engagement in online learning had become a large part of the academic community participation on a daily basis in the last period with the lockdown breakout, and e-Learning is becoming the innovative method to communicate between the universities and students. Last year, student at all levels struggled to adapt themselves to the change from traditional learning to virtual classes [6]. Despite its advantages, the main question in E-learning remains how ready students are for online learning, are they willing to accept it as a discipline even in the future when everything is back to normal? Researchers found that students' computer and internet efficacy and personal characteristic such as gender, ethnicity, course year level are significant difference in students' e-learning readiness and approach to e-learning [7]. They found a clear relationship between the level of education and technology use and adoption. Moreover, previous research found a clear relationship between the level of education and technology use and adoption [8].

2.1   Technological Complexity (TC) and Access to Technology

Technological Complexity is related to the user' perceived degree of difficulty in using an information system [9]. If the user will perform efforts to use a technology, their willingness and their motivation will be negatively affected [9, 10]. The use of computer networks, to support educational processes is becoming an actual field of interest and researchers have now been more studying and experimenting a number of factors influencing the educational approaches.

Financial barriers can also be an important factor that impacts students' access to technologies that shall be used to have advantage from learning online [11]. One of the main reason that might influence the students' dissatisfaction with online learning can be technological difficulties such as telecommunication infrastructure, the lack of internet access and consumption that can be break down during class session or online exams [11]. The accessibility to digital technologies due to technological problems such as connectivity and equipment, the expense of Internet connectivity can be a major problem that students shall face and can present a barrier to their engagement on e-learning [12].

2.2 Facilitating Conditions

Facilitating conditions are related to consumers' perceptions about available resources and facilities to perform a behavior [13]. Facilitating conditions in e-learning are the individuals' perceptions about technical and organizational infrastructure required to adopt a system using new technologies [14]. It is considered as one of the environmental factors and the external constraints that affect users' intention to adopt e-learning and their perception about how easy or difficult using this task. It is an extrinsic motivator for users to adopt e-learning [15]. The implementation of e-learning contains challenges like financing, skills, capacity to use it specially in developing countries where technological challenges, course challenges and context challenges take place [16]. Students may face a lack of competent people to manage and use e-learning systems during their transition from traditional learning to learning online. In the same context, institutions may also have a lack in identifying and developing staff to support whether technical support coupled with pedagogical skills to follow closely the users [11].

2.3 Learning Value

The construct «Price value» had been used by Venkatesh et al. (2012) [17] to identify the monetary costs and benefits associated with technology use. For Venkatesh (2012) [17] it is about the perceived monetary benefits for using the technology. The benefits of the learning and the value perceived became a priority objective for university research in the pandemic crisis [18]. The aim of e-learning is the effectiveness in terms of the learning outcomes. Moreover, benefits are stated in terms of cost efficiency and flexibility in time and place [19]. Transferring to virtual environment platform can provide students with free exchange of information, access to lectures and presentations with no costs or travel beside webinars [20]. The negative side of using e-learning is losing the social interaction and the learners' communication skills, and the stakeholders' involvement [21].

3   Methodology and Results

A. Participants

The survey's participants were students from a university in Lebanon situated in the capital Beirut. Saint Joseph University is a leading highly ranked and very well-established university in Lebanon for more than 146 years. The survey was precisely addressed to all 683 students in the business faculty. In this study, 458 responses were collected from the students using the google forms questionnaire in English. The questionnaire link was sent to students with the help of faculty members, it was anonymous and approved by the ethic committee of the university (USJ-2020-100). All the participants were regular students with high experience in E-learning system as a result of Covid-19 pandemic.

The questionnaire was divided into different sections: for demographics, participants had to choose between a set of predetermined answers and as for models' analysis, all variables were on a 7-point Likert scale (1: Strongly Disagree, 7: Strongly Agree).

B. Data Analysis

The data was filtered, cleaned and coded where SPSS v24.0 was the software adopted for dataset analysis. also records with missing data points were removed from the dataset. Ultimately our dataset was reduced to N = 445.

Demographic Statistics: 56% of the respondents (N = 445) were female participants, and the remaining 44% were males. The majority of the participants (50.7%) were between the age of 18–20. And 41.3% were between 21 and 23 years old. 82.5% percent of the participants were students in BA levels whereas the rest 17.5% were graduate students enrolled in the master program (Table 1).

To examine the variables under study among the participants, the variables were computed by taking mean of respective items and their mean (M), and standard deviations (SD) were: access to technology (2.2539 ± .577), learning value (3.8764 ± 1.746), facilitating conditions (4.1596 ± 1.477) and behavioral intentions (3.7663 ± 1.772) were. Table 2 shows the results of means medians and standard deviation for these variables.

The participants have good facilitating conditions (Mean 4.1596 SD 1.477) and learning value whereas they have a very low mean in terms of access to technology (Mean 2.2539 SD 0.577) which can be explained by the weak internet and WIFI accessibility in different regions in the country.

Table 1. Demographic statistics

		Frequency	Percentages %
Gender	*Female*	249	56
	Male	196	44
Age	*From 18 to 20 years*	226	50.7
	From 21 to 22 years	184	41.3
	23 years and more	36	8
Level of studies	*BA*	368	82.5
	MBA	77	17.5
	Total	445	100

As for the set of items in the computed variables, the researchers conducted a reliability test which results are shown in Table 3. Reliability coefficient (α) is the highest for the learning value (0.921) whereas it is the lowest for the access to technology (0.672). As for the facilitating conditions variable ($\alpha = 0.792$) and for behavioral intentions to use E-learning ($\alpha = 0.897$). All acceptable Cronbach alpha values.

*RQ1:Is there any demographic differences in terms of **Gender and learning levels**leading to noticeable disparities among these variables and the intention to use e-Learning in the future?*

Table 2. Mean, median and std. deviation

	Acces to technology	Learning value	Facilitating conditions	Behavioral intention
Mean	2.2539	3.8764	4.1596	3.7663
Median	2.00	4.00	4.00	4.00
Std. dev	0.57795	1.74643	1.47782	1.77298

Table 3. Reliability results

	Reliability statistics	
	Cronbach's alpha	N of items
Facilitating conditions	0.792	4
Access to technology	0.672	5
Learning value	0.921	4
Behavioral intentions to use E-learning	0.897	3

To determine the demographic differences, the authors ran Mann-Whitney U test.

The Mann-Whitney U test, also known as the Wilcoxon rank sum test, tests for differences between two groups on a single, ordinal variable with no specific distribution [22, 23].

The test's results are shown in Table 4 and 5.

a. Gender differences and acceptance of on-line learning:

H0: The students 'gender and learning levels have no effect on their behavioral intentions to use e-learning and to their learning value, their gender has no effect to their access to technology and to the on-line facilitating conditions.

H1: The students 'gender and learning levels have a positive effect on their behavioral intentions to use e-learning and to their learning value, their gender has positive effect to their access to technology and to the on-line facilitating conditions.

To determine the gender differences and learning levels, the Mann-Whitney U test results indicated that for all the variables tested in comparison first: to gender (0 female and 1 male), there are no gender differences for on line learning. Therefore, since Z score are all negative (below the mean) and a 2-tailed p-value higher than .05. This would normally be considered a non-significant result. We can be confident in accepting the null hypothesis that claims there are no demographic differences in online learning behavioral intentions and learning value and that man and women have same facilitating conditions and access to technology. Or, to put this another way, the result of the Mann-Whitney U test supports the proposition that students have no differences in e-Learning behaviors related to their gender. Furthermore, it was noted that in case of behavioral intentions, male students have the highest mean (225.73) compared to female students whereas the later have the highest means for the other three variables.

Table 4. Means

Level of studies		N	Mean rank	Gender		N	Mean rank
Acces to technology	BA	368	**223.34**	Acces to technology	Female	249	**226.75**
	MBA	77	221.38		Male	196	218.23
Facilitating conditions	BA	368	220.99	Facilitating conditions	Female	249	**224.40**
	MBA	77	**232.60**		Male	196	221.22
Learning value	BA	368	217.32	Learning value	Female	249	**224.20**
	MBA	77	**250.12**		Male	196	221.47
Behavioral intention	BA	368	218.62	Behavioral intention	Female	249	220.85
	MBA	77	**243.92**		Male	196	**225.73**
	Total	445			Total	445	

Table 5. Grouping variables

	Acces to technology	Facilitating conditions	Learning value	Behavioral influence
Grouping variable: level of studies				
Mann-Whitney U	14043.000	13429.000	12079.500	12557.000
Z	−0.141	−0.736	−2.065	−1.590
Asymp. Sig. (2-tailed)	0.888	0.462	0.039	0.112
Grouping variable: gender				
Mann-Whitney U	23468.000	24052.500	24103.000	23867.500
Z	−0.803	−0.265	−0.225	−0.402
Asymp. Sig. (2-tailed)	0.422	0.791	0.822	0.688

b. Level of studies differences and acceptance of on-line learning:
The Mann-Whitney U test results indicated that for all the variables tested in comparison to students' level of studies (1 BA and 2 MBA), there are no differences for on line learning except for the variable learning value where $Z = -2.065$ and p value $= 0.039$. Therefore, this would normally be considered as a significant result [23]. We can be confident in rejecting the null hypothesis that claims there is no level of studies differences in online learning towards the learning value and that BA and MBA students have different learning value. As for the other three variables we found that there are no differences among the students' behavioral intention or access to technology or the facilitating conditions. Furthermore, it was noted that in case of access to technology variable, BA students have the highest mean (223.34) compared to MBA students whereas the later have the highest means for the other three variables.

RQ2: *what is the relationship between access to technology, learning value perception and facilitating conditions leading to future behavioral intention to use eLearning?*

Spearman's correlation was calculated to examine the relationship between the four variables and to see the strength leading to behavioral intentions (model 1). Basically, a Spearman coefficient is a Pearson correlation coefficient calculated with the ranks of the values of each of the 2 variables instead of their actual values [24] (Fig. 1).

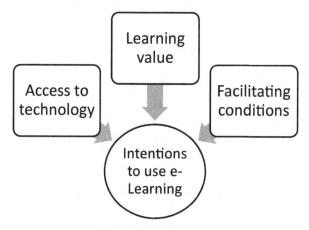

Fig. 1. Research model

There is an association between all the variables since they have all positive correlation with $p < 0.005$, the authors noted that the weakest variable is access to technology which relation to all the other three variables is weak. Whereas, the relationship between learning value and the behavioral intention is $r = 0.795$ the highest amongst them. Table 6 provides the R and R2 values (Table 7).

Table 6. Spearman's rho

		1	2	3	4
Spearman's rho	Acces to technology	1.000			
	Learning value	.364**	1.000		
	Facilitating conditions	.454**	.731**	1.000	
	Behavioral intention	.339**	.795**	.715**	1.000

Furthermore, a regression analyses was conducted to test our model.

The R value is 0.822 which indicates a high degree of correlation. The R2 value indicates how much of the total variation in the dependent variable, behavioral intention to use e-Learning, can be explained by the independent variables. In this case, 67.4%

Table 7. R square value

Model	R	R Square	Adjusted R square	Std. error of the estimate
1	.822[a]	0.676	0.674	1.01211

a. Predictors: (Constant), facilitating conditions, access to technology, learning value

can be explained, which is very large. In addition, the ANOVA which reports how well the regression equation fits the data (Table 8).

Table 8. Dependent variable

Model		Sum of Squares	df	Mean Square	F	Sig
1	Regression	943.952	3	314.651	307.168	.000[b]
	Residual	451.742	441	1.024		
	Total	1395.694	444			

a. Dependent Variable: behavioral intention
b. Predictors: (Constant), facilitating conditions, access to technology, learning value

It indicated that the regression model predicts the dependent variable significantly well. The statistical significance of the regression model that was run is $p < 0.0005$, which is less than 0.05, and indicates that, overall, the regression model is statistically significant. However, the coefficients table provides the researchers with additional necessary information on the independent variables that have the most and also least effect on the behavioral intention dependent variable. As a matter of fact, the researchers realized that access to technology variable with a $p = 0.67$ has no substantial influence on the behavioral intention to use e-Learning in the future (Table 9).

Table 9. Regression model results

Model		Unstandardized coefficients		Standardized coefficients	t	Sig
		B	Std. Error	Beta		
1	(Constant)	0.079	0.201		0.391	0.696
	Acces to technology	−0.040	0.094	−0.013	−0.426	0.670
	Learning value	0.595	0.041	0.586	14.611	0.000
	Facilitating conditions	0.354	0.051	0.295	6.969	0.000

a. Dependent Variable: behavioral intention

4 Discussion

Since a year for now, university students all around the world have been under discontinuous lockdowns due to the pandemic crisis COVID-19. Education lockdowns have contributed in growing the use of virtual learning environment and platforms within universities. Universities have confronted many challenges and obstacles for the assurance of quality virtual education, starting from technical problems to pedagogical hurdles. The main objective was always to test students' satisfaction in using e-learning and adopt it as transformational education system in the future even after the post pandemic era. Therefore, it was so crucial to propose this research paper theme for further analysis. The results of this study have revealed that facilitating conditions and learning value have positive impact on e-learning leading to acceptance of online engagement in the future at all times and not only in times of COVID-19.

E-learning technology is unique and represents a new era of distance education [25] but students may be exposed to a shortage of competent people to facilitate and guide them through the use of e-learning systems during transitions, which was the case during this critical period of pandemic.

However, the access to technology variable could be considered as the weakest variable in the model. The students lacking access to internet connectivity due to poor infrastructure and weak financial means in multiple developing countries and in the context of Lebanon as well. The price of Internet is expensive especially if Lebanese students have to pay a high bandwidth to fulfill online engagement [10]. Students might be dissatisfied with online learning due to technological difficulties where the infrastructure in Lebanon is proven to be below average with a bad connection preventing students from a normal online class. Sometimes students can be confronting problems in connectivity and lose data during online class sessions and they are obliged to leave then to re-log into session. It is same situation during online exam. Reports confirmed that Internet consumption and accessibility to digital technology and its expenses are major problems that students can confront [11], which is the case in Lebanon. Empirically knowing that the other two independent variables that are strongly affecting our model and influencing our dependent variable that is" the behavioral intentions to use online learning in the future", universities should invest in more facilitating conditions and work on the development of learning values through extensive internal and external campaigns that can secure students mobilization in this regard.

As for the demographic differences, no gender difference has been revealed on the impact of accepting and using e-learning for instance. This is due to the equality of gender when accessing e-learning environment which is normal. Both male and female can access the digital platforms and have the same learning. On the other hand, the results specified a variation in learning values with respect to levels of education whereas students of the graduate programs scored higher values than the ones in bachelor, therefore, BA students should be exposed to extensive awareness and informative campaigns on the advantages of virtual learning and its impact on their educational and professional life.

5   Limitations and Conclusion

This study was conducted on more than 65% of the students of the faculty of business within the same university which is considered as a big sample. The fact of enlarging the survey to include more students from other faculties and universities could give a wider perspective into the research. But as lockdowns were considered a threat for our survey administration, inflating the sample survey was a serious challenge. Also, it could be worthwhile considering the implication of online learning on sustainable environment and its related contexts and framework in relation to the University students. As we mentioned before, technical and pedagogical support from institutions can enhance their relationship with students [10]. Concerning this situation, the study could be widened to reveal if students have a good pedagogical and organizational support from their institutions in terms of administrative works in the times of COVID 19 leading to their unsatisfaction from the online experience in learning.

This research paper has tackled on some constituents that have solid effect on students' intention to use online learning in the future where researchers found no differences in gender but a major difference in accepting and using on-line education for different level of education. Moreover, the importance in the area of innovation in the curricula of Business School degrees in a developing country was examined in the same business school [26], while regarding innovation and focusing on employability and education and actions to be taken in the coming years by higher education institutions in order to perform better while maintaining their mission and reaching their goals. In addition, the authors research results relating to the revision of master's programs leads to innovate in the curriculum and shows that by adding some online courses that are mainly accepted by master's students that would be interesting and innovating in the future and leading to a hybrid modality that would also facilitate the employability and availability of students to join classes at home after long working days.

Due to COVID-19, learning and styles in academic institutions around the world have been forced to adopt the use of online teaching reinforced with new technical and pedagogical methods to secure quality delivery of courses through cyber space.

This leap in the methods have impeded students with load of personal work and effort to align with the stream. Such intensive effort and students' engagement could have led them to more psychological disruption followed with anxiety, stress and negative affect, a substance phenomenon that could be subject to deeper quantitative analysis in the future. In addition to this, also a qualitative research could be complimentary for a profound evaluation of the student's adaptation and integration challenges when using and adopting online technology through different platforms.

References

1. Crawford, J., Butler-Henderson, K., Rudolph, J., Malkawi, B., Glowatz, M., Burton, R.: COVID-19: 20 countries' higher education intra-period digital pedagogy responses. J. Appl. Learn. Teach. **3**, 1–20 (2020)
2. Nuere, S., de Miguel, L.: The digital/technological connection with COVID-19: an unprecedented challenge in University teaching. Techn. Knowl. Learn. (2020). https://doi.org/10. 1007/s10758-020-09454-6

3. Becker, K., Newton, C., Sawang, S.: A learner perspective on barriers to e-learning. Aust. J. Adult Learn. **53**(2), 35–57 (2013)
4. Jones, C., Ramanau, R., Cross, S., Healing, G.: Net generation or digital natives: is there a distinct new generation entering university? Comput. Educ. **54**, 722–732 (2010)
5. Shahnavazi, A., Mehraeen, E., Bagheri, S., Miri, Z., Mohammadghasemi, M.: Survey of students readiness to use of e-learning technology. J. Paramed. Sci. Rehabil. **6**, 60–66 (2017)
6. Sanchez-Gordon, S., Luján-Mora, S.: MOOCs gone wild. In: Proceedings of the 8th International Technology, Education and Development Conference (INTED 2014), Valencia (Spain), 10–12 March 2014, pp. 1449–1458 (2014). ISBN 978-84-616-8412-0
7. Lau, C.Y., Shaikh, J.: The impacts of personal qualities on online learning readiness at Curtin Sarawak Malaysia (CSM). Educ. Res. Rev. **7**(20), 430–444 (2012)
8. Riddell, C., Song, X.: The role of education in technology use and adoption: evidence from the Canadian workplace and employee survey. Sage J. **70**(5) (2017)
9. Teo, T., Huang, F., Ka, C., Hoi, W.: Explicating the influences that explain intention to use technology among English teachers in China. Interact. Learn. Environ. **26**(4), 460–475 (2017). https://doi.org/10.1080/10494820.2017.1341940
10. Goh, C.F., Hii, P.K., Tan, O.K., Rasli, A.: Why do university teachers use e-learning systems? Int. Rev. Res. Open Distrib. Learn. **21**(2), 136–155 (2020)
11. Fawaz, M., Samaha, A.: E-learning: depression, anxiety, and stress symptomatology among Lebanese university students during COVID-19 quarantine. Nurs Forum **56**, 52–57 (2021)
12. Olaniran, B., Agnello, M.: Globalization, educational hegemony, and higher education. Multicultural Educ. Technol. J. **2**(2), 68–86 (2008)
13. Brown, S., Venkatesh, V.: Model of adoption of technology in households: a baseline model test and extension incorporating household life cycle. MIS Q. **29**(3), 399–436 (2005)
14. Al-Hujran, O., Al-Lozi, E., Al-Debei, M.: "Get ready to mobile learning": examining factors affecting college students' behavioral intentions to use m-learning in Saudi Arabia. Int. J. Bus. Adm. **10**(1), 16 (2014)
15. Kasse, J.P., Moya, M., Nansubuga, A.: Facilitating condition for e-learning adoption—case of Ugandan universities. J. Commun. Comput. **12**, 244–249 (2015)
16. Josephat, O.O., Herbert, W., Ngumbuke, F.: Challenges of e-learning in developing countries: the Ugandan experience. In: Proceedings of INTED2012 Conference (2012)
17. Venkatesh, V., Thong, J., Xu, X.: Consumer acceptance and use of information technology: extending the unified theory of acceptance and use of technology. MIS Q. **36**(1), 157–178 (2012). Accessed 4 Apr 2021
18. Kamarianos, I., Adamopoulou, A., Lambropoulos, H., Stamelos, G.: Towards an understanding of university students' response in times of pandemic crisis (COVID-19). Eur. J. Educ. Stud. **7**(7) (2020)
19. Davis, R., Wong, D.: Conceptualizing and measuring the optimal experience of the elearning environment. Decis. Sci. **5**(1), 97–126 (2007)
20. Shah, S., Diwan, S., Kohan, L., Rosenblum, D., Gharibo, C., Soin, A., et al.: The technological impact of COVID-19 on the future of education and health care delivery. Pain Phys. **23**, 367–380 (2020)
21. Al-Qahtani, A., Higgins, S.E.: Effects of traditional, blended and e-learning on students' achievement in higher education. J. Comput. Assist. Learn. **29**(3), 220–234 (2013)
22. Mann, H.B., Whitney, D.R.: On a test of whether one of two random variables is stochastically larger than the other. Ann. Math. Statist. **8**(1), 50–60 (1947)
23. MacKnight, P., Najab, J.: Mann-Whitney U Test. Mathematics (2010)
24. Kutner, M.H., Nachtsheim, C.J., Neter, J., Li, W.: Inferences in regression and correlation analysis. In: Applied Linear Statistical Models (International Edition), 5th edn., pp. 40–99. McGraw-Hill/Irvin, Singapore (2005)

25. Garrison, D.R.: Computer conferencing: the post-industrial age of distance education. Open Learn. **12**(2), 3–11 (1997)
26. Aoun, G., Mallah Boustani, N.: Impact of innovation on master programs in business schools. In: INTED2017 Proceedings, pp. 1086–1094 (2017). https://doi.org/10.21125/inted.2017.0405

The Importance of Tacit Knowledge When Teaching Suddenly Online

Pierre-Emmanuel Arduin[1]([⊠]), Michel Grundstein[2], Brice Mayag[2], Elsa Negre[2],
Camille Rosenthal-Sabroux[2], and Inès Saad[3]

[1] Université Paris-Dauphine, PSL, CNRS DRM, Paris, France
`pierre-emmanuel.arduin@dauphine.psl.eu`
[2] Université Paris-Dauphine, PSL, CNRS LAMSADE, Paris, France
`{brice.mayag,elsa.negre}@dauphine.psl.eu`
`camille.rosenthal-sabroux@lamsade.dauphine.fr`
`michel.grundstein@yahoo.fr`
[3] Université de Picardie Jules Verne, MIS and Amiens Business School,
Amiens, France
`ines.saad@esc-amiens.com`

Abstract. The COVID-19 outbreak had multiple impacts on our lives. While the overall society has been impacted, teachers and students have not been spared: lockdowns and emergency remote teaching led them to adapt. So that teachers suddenly taught online to transfer knowledge. Besides such an adaptation, one of the challenges has been to keep students motivated and interested to ensure knowledge transfer, even if there were no physical interaction. Human responses are then crucial as answers to these challenges, when individuals are facing such a crisis situation. In this paper, we propose to model teaching mechanisms and how they have been impacted to suddenly go online and ensure knowledge transfer, whether tacit or made-explicit. We focus on the importance of tacit knowledge transfer from the teacher towards the students. Through our proposal we highlight the role of interactions between the teacher and the students and how much informal exchanges are essential to transfer tacit knowledge.

Keywords: Online teaching · Synchronous teaching · Knowledge transfer · Information and knowledge system

1 Introduction

While societies are still far from seeing all the consequences of the crisis caused by the COVID-19, some authors draw up a "balance sheet" concerning the sudden shift online of their classes in higher education [22]. The issue of knowledge sharing is crucial in higher education, remotely in general but online in particular. Such a question has been widely and for a long time discussed in the literature [12].

Current events have made it possible – and necessary – for knowledge to be suddenly shared online [3], a practice hitherto relying on the acceptance and the

© Springer Nature Switzerland AG 2021
I. Saad et al. (Eds.): ICIKS 2021, LNBIP 425, pp. 29–42, 2021.
https://doi.org/10.1007/978-3-030-85977-0_3

choice of people and cultures [18]. This raises the question of the impact that a crisis such as a global pandemic had, is having, and will have on the daily lives of teachers in higher education: recruiting students, transferring knowledge, and deciding the assessment of students. How could the mechanisms deployed by individuals in such cases be modeled? Notably as the underlying processes were face-to-face in the past and moved suddenly online, without a choice neither from the teachers nor from the students.

Global crisis, as the pandemic we are experiencing in these days are challenging the field of information and knowledge systems to help building effective management solutions and to support Humans keep on performing core activities as teaching. How models can help us to understand changes in our way to teach online nowadays? We propose to rely on the model of the Organization's Information and Knowledge System (OIKS) [19] as a thought structure to understand knowledge transfer, taking into account knowledge absorption, use and a human-centred approach. For authors such as [23], science requires boldness, in order to open our mind towards another thought that immediate thought in order to provoke, test and sometimes contest it. It is exactly what we want to do with the models proposed in this paper.

Our goal is to enhance the importance of tacit knowledge transfer from the teacher towards the students. So with the models we highlight the importance of the interactions between the teacher and the students. Informal exchanges are essential, because tacit knowledge is transferred.

In this paper, background theory and assumptions are first presented: a literature review on crisis and behavior, knowledge transfer, and the Organization's Information and Knowledge System (OIKS). Models representing teaching and knowledge transfer in face-to-face and online situations will then be exposed. Finally, we present some discussions, problems and limits arising. We conclude, by highlighting the role of tacit knowledge and provide the beginning of a response opening the way to a transfer of knowledge taking into account human capacities reinforced by the digital transformation.

2   Background Theory and Assumptions

In this section, we first present crisis and behavior. Second, we present knowledge transfer, drawing from the knowledge management literature. Third, we present the Organization's Information and Knowledge System (OIKS) as a way to consider the importance of tacit knowledge in knowledge transfer.

2.1   Crisis and Behavior

The correlation between a person's behavior and her/his context has been proven [28]. However, the notion of context remains vague. Indeed, due to a lack of consensus, there is still no single definition for the context. Context can influence a person's interests, which is why it is important to take into account this type of additional information [9]. Suppose that a young man of 20 who loves war

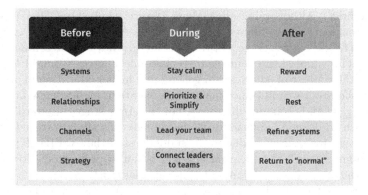

Fig. 1. Before, during and after a crisis (https://www.poppulo.com/blog/three-key-communication-lessons-for-managing-a-crisis)

or action films wants to go see a movie with a girlfriend. Instead of choosing a war or action movie, he will tend to choose a romantic movie. Here, the context influences the preferences, desires and interests of people and hence their decisions. The notion of context has been studied in various fields, notably in pervasive and ubiquitous computing but it is difficult to establish a standard (single) definition due to the multifaceted nature of the context [10]. The most accepted definition is that of [15] which defines context as any information that can be used to characterize the situation of an entity. An entity can be a person, place, or object that is considered relevant in the interaction between a user and an application, while including the latter two. In order to know the context, it is necessary to collect the contextual information. This context can be captured, collected explicitly or implicitly.

[32], based on a bibliographic study, proposed a hierarchical categorization of context factors including, in particular, demographic, cognitive, environmental and temporal information, etc. Moreover, the modeling of the context is still complicated given the nature of the data and/or contextual information: in fact, the model must be able to manage the various data sources, the heterogeneity in terms of their quality, lifespan, and their potential imperfect nature [21].

Before and during a crisis, people often act according to their own interpretative frameworks [31], which do not always allow them to react in an appropriate manner to risk situations and which may induce dangerous reactions. Behavior is a concept that needs to be clarified and well defined, it can be approached in very different ways within the scientific world. We take here the definition proposed by [30] for which the behavior corresponds to the reactions of an individual, considered in a milieu and in a given unit of time to an excitation or a set of stimulations.

Indeed, a crisis can be divided into three main stages: (i) before the crisis, (ii) during the crisis and (iii) after the crisis, as detailed in Fig. 1. The COVID-19 crisis is a multi-faceted crisis, different from the usual crises considered sudden (and often brief). Although initially sudden, the crisis linked to COVID-19 drags on and seems stuck in an infinite loop "During → After → During" ... Indeed, with regard to the *During* stage, "Stay calm" has given way to weariness, "Prioritize and simplify" and "Lead your team" remain unchanged. "Connect leaders to teams", in the context of education in general and higher education in particular is already showing its limits since many students are in an academic failure and in psychological and economic distresses. For the *After* stage, "Reward", "Rest" and "Return to normal" appear to be compromised as those involved attempt to "Refine systems". As such, due to the suddenness of the COVID-19 pandemic and its "infinite loop", the diagram of Fig. 1 is not appropriated. How to best approach the operating mode presented in this diagram?

The questions we ask in this paper, in connection with the COVID-19 outbreak, are in the last two stages, namely during and after the crisis. If decisions were made during the crisis, should they be maintained? If the crisis is "past", the return to normal in the case of a pandemic seems difficult, not to say impossible, how then to get a "pretend" normality? What behavior is adopted by Humans in such a particular context? How are the knowledge transfer processes impacted?

2.2   Knowledge Transfer

Knowledge is a large and abstract notion which brings an abundant literature about it [16]. However, the definitions of knowledge presented till now in the literature do not allow a consensus on what is knowledge and how to define it exactly. This engenders a certain confusion, in particular, by mixing together knowledge of the concepts of data and information. The clarification of these three concepts (knowledge, data, and information) is decisive for the study of knowledge transfer processes proposed in this paper. Indeed, different conceptions of knowledge imply different perceptions of the knowledge transfer process itself.

Data are symbols, facts, and raw numbers used to communicate information. They are devoid of context and they are not directly significant. Information is a set of data to which context has been added and given meaning by linking them together according to a specific presentation (textual, auditory, and/or visual). The information is therefore located at a context level that corresponds to the understanding of the relationships between the data.

Information in turn becomes knowledge when it is cognitively reappropriated by a recipient. As such, knowledge encompasses notably data, contextual information, and expertise [14]. Knowledge is distinguished from information according to certain aspects which depend in this case on the receiving subject. The receiver gives meaning to information according notably to his or her own beliefs and values, his or her socio-professional environment [29]. The interpretative framework [31] of the receiving subject allows him or her to construct a representation of information and to create his or her own knowledge. We distinguish two types of knowledge: Tacit and explicit. Tacit knowledge is embedded in actions, in experience and mobilized by the knowledge carrier in a specific context. It includes cognitive and technical elements. The cognitive component refers to individual's mental models, *i.e.* his or her beliefs and preferences. The technical component consists of technical know-how which applies to a specific context. Knowing how to convince a type of customer for making a sale is an example of tacit knowledge.

Previous works on knowledge transfer occur at different levels: individual, collective or organizational [4,7,11,17].

In this paper, we focus only on the knowledge transfer at the individual level which means the knowledge transfer between a teacher and a student. Mainly conceptual frameworks [4,7] have been proposed to study and identify the most important elements in the process of knowledge transfer. Among these elements, we cite the three determinant elements in our context that is related to a sudden change in the teaching approach from face-to-face to online: (1) the motivation disposition of the source of knowledge (the teacher in our context) to transmit his or her knowledge; (2) the motivation of the receiver (the student) to acquire knowledge from the source and (3) the absorptive capacity of the receiver of knowledge to use the knowledge and not only to acquire and understand it [14]. Knowledge is thus transferred when information is transmitted by a sender and then this information is absorbed and interpreted into tacit knowledge based on the interpretative framework of the receiver and finally used. The transmission of knowledge is necessary but not sufficient to speak of a process of transfer. Knowledge can only be transferred if it is absorbed and used. This definition strongly depends on the absorptive capacity of the receiver which reflects its capacity to assimilate and apply knowledge [13]. This raises then the importance of a model allowing to consider individuals, information and knowledge as a system within organizations.

2.3   The Organization's Information and Knowledge System (OIKS)

Information and knowledge transfer have a direct impact on the functioning of organizations [20]. Such an impact increases when companies are border-less, open and adaptive, called "extended companies". First proposed in 2009 [20] and instantiated in 2015 [5], the Enterprise's Information and Knowledge System (EIKS) bridges domains such as Knowledge Management and Information Systems. In this paper, we use an extension of the EIKS to general organizations such as Universities: the Organization's Information and Knowledge System (OIKS) [19].

The OIKS rests on a socio-technical context, which consists of individuals in interaction among them, with machines, and with the OIKS itself as illustrated by Fig. 2 [19]. It includes:

- An *Information System* (IS), made up of individuals and digital artifacts, all of which are "processors of information";
- A *Knowledge System* (KS), made up of both tacit knowledge embodied by individuals and made-explicit knowledge that has been formalized on an appropriate support medium (documents, videos, photos, etc.);

– A *Digital Information System* (DIS), an artifact constructed relying on digital technology.

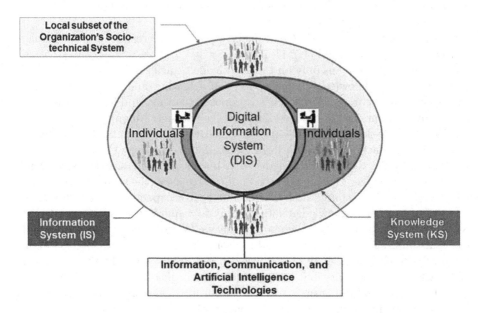

Fig. 2. The Organization's Information and Knowledge System (OIKS)

The OIKS highlights the importance to distinguish data, information, tacit and made-explicit knowledge. Rather than knowledge, the IS carries *information as a source of knowledge*. Only individuals can possess genuine knowledge, resulting from their interpretation of information [31] as presented Sect. 2.2. We argue that individuals are carriers of knowledge and processors of information. They are a component of the OIKS, whose behavior will be modeled when suddenly teaching online in Sect. 3.

Mirrored with the ISO/IEC 30401:2018 standard (Knowledge management systems—Requirements), the OIKS underlines the importance of individuals and their interpretative processes: absorbing technology is clearly showed as insufficient to transfer knowledge. Modeling human responses for knowledge transfer in face-to-face and online situations appears then as a way to highlight the importance of tacit knowledge during teaching activities.

3   Modeling Human Responses for Knowledge Transfer

It cannot be neglected that the COVID-19 outbreak has affected the mental health of students [25]. Some of them deployed strategies observed notably by [8] not only to transfer their knowledge but also to reduce the effect of anxiety during the sudden switch online of courses.

Knowledge transfer can be characterized as "Transfer = Transmission + Absorption (and Use)" [14], as presented Sect. 2.2. However there is a lack in the literature of works about tacit knowledge and its transfer, particularly when bridging knowledge management and information systems to model such a knowledge transfer process [6].

Teaching in general and higher education in particular implies preparing a lesson, evaluating and communicating with students. Teaching may be synchronous and/or asynchronous. It may be face-to-face or online through the access to digital tools, as well as hybrid. Knowledge may be transferred during teaching. Figure 3 presents a use case diagram showing these items and their relations. In this paper, we focus on the case of synchronously teaching in higher education, whether face-to-face or online.

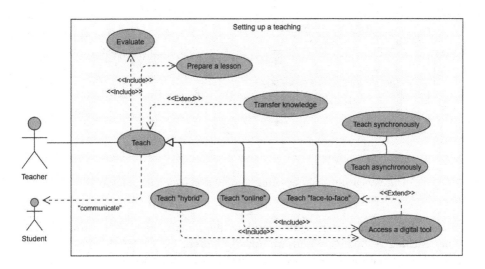

Fig. 3. Setting up a teaching: Use case diagram

In this section, we first present two situations of teaching using UML (Unified Modeling Language[1]) diagrams: the face-to-face and online situations. Second, we focus on the knowledge transfer itself, again by comparing the face-to-face and online situations. The proposed models highlight the importance of tacit knowledge within teaching situations and the existence of possible undetected misunderstandings in online situations that jeopardize tacit knowledge transfer.

3.1 Face-to-Face Teaching and Online Teaching

We remind that the models presented in this paper represent the case of teaching in higher education, whether in a University or a Business School. In the following the words "teaching" or "teacher" refers to the case of higher education.

---
[1] https://www.omg.org/spec/UML.

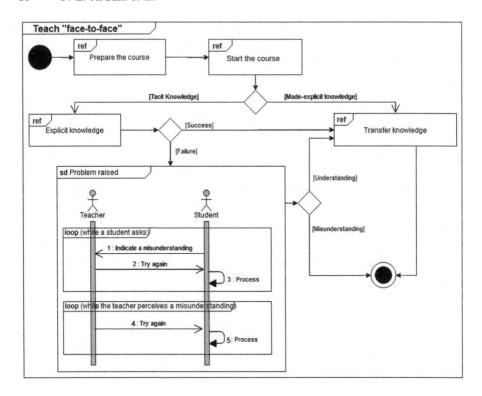

Fig. 4. Face-to-face teaching: Interaction overview diagram

Figures 4 and 5 show interaction overview diagrams modeling face-to-face and online teaching respectively and are now presented.

Teaching has always been a face-to-face activity. The sentence "I hear and I forget, I see and I remember, I do and I understand", attributed to Confucius, clearly illustrates the idea of learning by doing that has been formalized in the 1980s [24]. When teaching face-to-face, a teacher has to prepare and start a course. Such knowledge may have already been made explicit on supports such as documents, videos, photos, etc. It may then be transferred as it is (ref ''Transfer knowledge'' in Fig. 4, detailed in Sect. 3.2). When tacit, knowledge has to be made explicit during the course (ref ''Explicit knowledge'' in Fig. 4), and such process may success or fail. Failure raises a problem between the teacher and a student misunderstanding (sd ''Problem raised'' in Fig. 4). Such a misunderstanding may be indicated by the student, but also – and most often as teachers would have experienced – may be perceived by the teacher during the course.

Teaching online is a particular case of remote teaching, whose first evidence may be found in 1728 with advertisements in the Boston Gazette for shorthand lessons by mail [26]. When teaching online, a teacher has to prepare a course as in the face-to-face situation. He/she accesses a digital tool to create a virtual

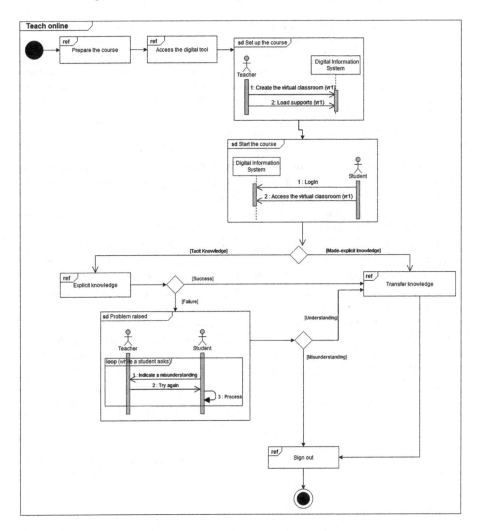

Fig. 5. Online teaching: Interaction overview diagram

classroom and loads supports (sd ''Set up the course'' in Fig. 5). At the beginning of the course, students have to access such a virtual classroom. As in the face-to-face situation, knowledge is then transferred when it is already made-explicit (ref ''Transfer knowledge'' in Fig. 5, detailed in Sect. 3.2), and it has to be made explicit during the course when it is tacit to be actually transferred (ref ''Explicit knowledge'' in Fig. 5). Such process of making explicit tacit knowledge may fail and the reader would have noticed that compared to the face-to-face situation, misunderstanding can only be indicated by students in the online situation (sd ''Problem raised'' in Fig. 4).

3.2   Knowledge Transfer When Teaching

In the previous section (Sect. 3.1) we presented interaction overview diagrams (Figs. 4 and 5) modeling teaching in higher education. In this section, we focus on a specific part of such a teaching activity: knowledge transfer. Figure 6.a presents an interaction overview diagram of such an activity in the face-to-face situation and Fig. 6.b in the online situation.

Knowledge transfer implies two or more actors (see discussion, Sect. 3.3). When the receiver tries to interpret knowledge by processing it (`self messages 2` in Figs. 6.a and 6.b), he/she can create knowledge by absorbing it (`self message 9` in Fig. 6.a and `self message 6` in Fig. 6.b). Otherwise, a misunderstanding may occur. If the face-to-face situation allows the teacher to observe and detect such a misunderstanding (`message 6` in Fig. 6.a), the reader would have noticed that the online situation deprives the teacher of such an ability.

This posture may remind in a moderate manner the probationers (ἀκουσματικοί, the auditors, those who can only listen) in the school of Pythagoras. These students had to listen to lectures given by Pythagoras behind a curtain during five years by remaining silent. In this case neither the teacher nor the students would communicate on misunderstandings.

The online situation deprives the teacher not only of the ability to observe students misunderstanding, but also – and maybe more depressing for the teacher – the ability to appreciate that students understood (`message 10` in Fig. 6.a). This leads us to discuss the limits and perspectives of this research.

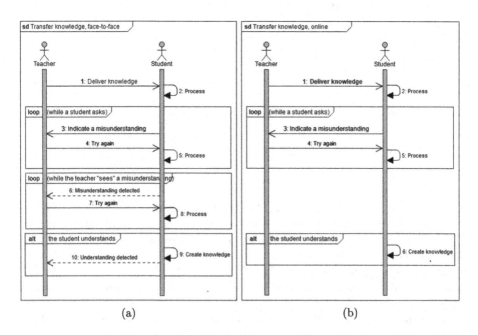

Fig. 6. Face-to-face (a) and online (b) knowledge transfer: Interaction overview diagrams showing limited misunderstanding perception possibilities for the teacher in the online situation

3.3   Discussions, Problems and Limits

Authors such as [27] are interested in measuring a degree of user satisfaction from the point of view of students (factors influencing the preference of online teaching for example). [2] analyzed the results of a survey from French students and showed that they would like to keep 35% of their courses on-line after experiencing a lockdown. This percentage could underline the fact that some students, often shy during classic lectures, feel more comfortable during an online teaching. [1] made an experiment with innovative features during a distance summer school and noted an expansion of participants' knowledge, as well as significant financial benefits for the organizers.

To check whether the transfer of tacit knowledge, from the teacher to the students, has been carried out, it seems necessary to evaluate the students at the end of the lessons. However, evaluating online students, in a sudden crisis, requires to develop new ways in order they are evaluated and monitored, following the rules of equality defined by the administration. Indeed, the assessment of students, initially performed in face-to-face, is likely to raise the question of diplomas level. For instance, in France, some universities, like the University of Rennes 1, have considered monitoring exams via the student webcam for example. Some authors like [33] have also proposed approaches using an online learning system (*Active Learning System*) based on smartphones to monitor students, raising the question of security and privacy. If the models proposed in this paper highlight how lacks in tacit knowledge transfer or undetected misunderstandings lead to problematic situations for students assessments, we are currently extending them to integrate this question of evaluations. We may think that the implementation of these evaluation procedures is an additional difficulty for teachers, especially since they are sometimes psychologically affected by the lack of real interactions with their students. Therefore problems arising during the assessment of students in a teaching during online classes in a sudden crisis can be viewed as a limitation of such a model of teaching, as well as the need to consider interactions between students in future works.

4   Conclusions and Perspectives

In this paper, background theories and assumptions were presented, the context of crisis and behaviors it induces were discussed as well as knowledge transfer processes and the importance of a model considering individual within organizations. We highlighted and described, in the form of a UML representation, the face-to-face teaching processes and synchronous online teaching processes. In addition, we addressed the issue of knowledge transfer by relying on the OIKS model on which our vision of the organization's information and knowledge system is based. We used the example of online teaching in higher education, exemplifying the role of tacit knowledge of teachers and students. Finally, we mentioned some of the problems that arise when assessing students in the case of online teaching caused by a sudden health crisis, as well as perspectives for future works.

Taking the OIKS model into account allows us to highlight the importance of the role of tacit knowledge in knowledge transfer between a teacher and a student. Similarly, we find that an individual's tacit knowledge impacts his or her behavior. Indeed, we believe that an individual's behavior is triggered by information that, in a given context and situation, is interpreted through the own interpretative frameworks of the individual. Information is then transformed into tacit knowledge which makes the individual act in the form of interactions with others and with the digital information system. In the OIKS model, individuals and their interactions with the digital information system are represented. These interactions result in the transfer of explicit knowledge and tacit knowledge that has been made explicit. We remark that non explicitable tacit knowledge can be demonstrated by video or other means (virtual reality for example). Human responses for knowledge transfer are considered and it is in this sense that the OIKS model remains a reference in our thinking scheme.

Generally speaking, the crisis generated by the COVID-19 induces constructivist behaviors - the path is made by walking - in all areas of human activities while they were deterministic. Informal time in face-to-face is irreplaceable but what about when teaching online that becomes inevitable in the form of hybrid teaching? By highlighting the role of tacit knowledge, this paper provides the beginning of a response opening the way to a transfer of knowledge taking into account human capacities reinforced by the digital transformation and the numerous applications it makes available to us.

References

1. Abeyweera, R., Guillerme, G., Senanayake, N.S., Peiris, A., Jayasuriya, J., Fransson, T.H.: Dissemination of knowledge and sharing experiences in emerging issues through a remote summer school. In: 2019 IEEE Global Engineering Education Conference (EDUCON), pp. 152–159. IEEE (2019)
2. Adam-Ledunois, S., Arduin, P.E., David, A., Parguel, B.: Basculer soudain aux cours en ligne : le regard des étudiants. In: Le Libellio d'Aegis. vol. 16 - Série spéciale Coronam, semaine 5, pp. 55–67 (2020)
3. Al Lily, A.E., Ismail, A.F., Abunasser, F.M., Alqahtani, R.H.A.: Distance education as a response to pandemics: coronavirus and Arab culture. Technol. Soc. 101317 (2020)
4. Alavi, M., Leidner, D.E.: Knowledge management and knowledge management systems: conceptual foundations and research issues. MIS Q. **25**(1), 107–136 (2001)
5. Arduin, P.E., Grundstein, M., Rosenthal-Sabroux, C.: Information and Knowledge System. Wiley-ISTE (2015)
6. Arduin, P.E., Rosenthal-Sabroux, C., Grundstein, M.: Considering tacit knowledge when bridging knowledge management and information systems for collaborative decision-making. Inf. Syst. Knowl. Manag. 131–158 (2014)
7. Argote, L., Erin, F.: Knowledge transfer in organizations: the roles of members, tasks, tools, and networks. Organ. Behav. Hum. Decis. Process. **136**(1), 146–159 (2016)
8. Baloran, E.T.: Knowledge, attitudes, anxiety, and coping strategies of students during COVID-19 pandemic. J. Loss Trauma 1–8 (2020)

9. Baltrunas, L., Ludwig, B., Peer, S., Ricci, F.: Context relevance assessment and exploitation in mobile recommender systems. Pers. Ubiquit. Comput. - PUC **16**, 1–20 (2012). https://doi.org/10.1007/s00779-011-0417-x
10. Bazire, M., Brézillon, P.: Understanding context before using it. In: Dey, A., Kokinov, B., Leake, D., Turner, R. (eds.) CONTEXT 2005. LNCS (LNAI), vol. 3554, pp. 29–40. Springer, Heidelberg (2005). https://doi.org/10.1007/11508373_3
11. Bouzayane, S., Saad, I.: Prediction method based DRSA to improve the individual knowledge appropriation in a collaborative learning environment: case of MOOCs. In: 50th Hawaii International Conference on System Sciences (HICSS), pp. 124–133 (2017)
12. Burian, S., Horsburgh, J., Rosenberg, D., Ames, D., Hunter, L., Strong, C.: Using interactive video conferencing for multi-institution, team-teaching. In: Proceedings of the ASEE Annual Conference and Exposition, vol. 23, p. 1 (2013)
13. Cohen, W.M., Levinthal, D.A.: Absorptive capacity: a new perspective on learning and innovation. Adm. Sci. Q. **35**(1), 128–152 (1990)
14. Davenport, T.H., Prusak, L., et al.: Working Knowledge: How Organizations Manage What They Know. Harvard Business Press (1998)
15. Dey, A.: Understanding and using context. Pers. Ubiquit. Comput. **5**, 4–7 (2001). https://doi.org/10.1007/s007790170019
16. Fantl, J.: The Philosophy of Information. Oxford University Press, Oxford (2012)
17. Ghrab, S., Saad, I., Kassel, G., Gargouri, F.: A core ontology of know-how and knowing-that for improving knowledge sharing and decision making in the digital age. J. Decis. Syst. **26**(2), 138–151 (2017)
18. Grippa, F., Secundo, G.: How km 2.0 supports remote cross-cultural learning communities. In: Proceedings of the 9th European Conference on Knowledge Management: ECKM2008, p. 241. Academic Conferences Limited (2008)
19. Grundstein, M.: Toward management based on knowledge. In: Current Issues in Knowledge Management. IntechOpen (2019)
20. Grundstein, M., Rosenthal-Sabroux, C.: Vers une approche du système d'information et de connaissance transposée de l'approche du knowledge management dans l'entreprise étendue. In: Management et gouvernance des SI, pp. 85–127. Hermès Science - Lavoisier (2009)
21. Henricksen, K., Indulska, J.: Developing context-aware pervasive computing applications: models and approach. Pervasive Mob. Comput. **2**(1), 37–64 (2006). https://doi.org/10.1016/j.pmcj.2005.07.003
22. Huang, R., Tlili, A., Chang, T.W., Zhang, X., Nascimbeni, F., Burgos, D.: Disrupted classes, undisrupted learning during COVID-19 outbreak in china: application of open educational practices and resources. Smart Learn. Environ. **7**(1), 1–15 (2020)
23. Klein, É.: Le goût du vrai. Gallimard (2020)
24. Kolb, D.A.: Experience as the Source of Learning and Development. Prentice Hall, Upper Sadle River (1984)
25. Lee, J.: Mental health effects of school closures during COVID-19. Lancet Child Adolescent Health **4**(6), 421 (2020)
26. Moore, M.G.: Distance education: the foundations of effective practice. J. High. Educ. **63**(4), 468–472 (1992)
27. Nariman, D.: Impact of the interactive e-learning instructions on effectiveness of a programming course. In: Barolli, L., Poniszewska-Maranda, A., Enokido, T. (eds.) CISIS 2020. AISC, vol. 1194, pp. 588–597. Springer, Cham (2021). https://doi.org/10.1007/978-3-030-50454-0_61

28. Riboni, D., Bettini, C.: COSAR: hybrid reasoning for context-aware activity recognition. Pers. Ubiquit. Comput. **15**, 271–289 (2011). https://doi.org/10.1007/s00779-010-0331-7
29. Saad, I., Rosenthal-Sabroux, C., Gargouri, F.: Information Systems for Knowledge Management. Wiley, Hoboken (2014)
30. Sillamy, N.: Dictionnaire usuel de psychologie. Bordas (1983)
31. Tsuchiya, S.: Improving knowledge creation ability through organizational learning. In: Proceedings of the International Symposium on the Management of Industrial and Corporate Knowledge, ISMICK 1993, pp. 87–95 (1993)
32. Vahidi Ferdousi, Z., Negre, E., Colazzo, D.: Context factors in context-aware recommender systems. In: Atelier interdisciplinaire sur les systèmes de recommandation, AISR 2017, Paris, France, May 2017
33. Yamamoto, N.: An integrated online learning approach using a smartphone-based active learning system and a web video on-demand system. In: Barolli, L., Li, K.F., Enokido, T., Takizawa, M. (eds.) NBiS 2020. AISC, vol. 1264, pp. 369–374. Springer, Cham (2021). https://doi.org/10.1007/978-3-030-57811-4_35

Decision Support System for Online Recruitment

Halima Ramdani[1,2,3](✉), Davy Monticolo[1](✉), Armelle Brun[2](✉),
and Eric Bonjour[1](✉)

[1] Equipe de Recherche sur les Processus Innovatifs, Université de Lorraine, ERPI,
Nancy, France
{davy.monticolo,eric.bonjour}@univ-lorraine.fr
[2] Université de Lorraine, CNRS, Loria, Nancy, France
armelle.brun@loria.fr
[3] Xtramile, 11 rempart Saint-Thiebaut, 57000 Metz, France
h.ramdani@myxtramile.com

Abstract. In the past, potential candidates for a job offer were in physical locations that could be reached through the major media that were available at the time, often strongly rooted in their local geographic space. Today, digital media replaced those traditional channels, offering advertisers a broader geographic reach. However digital channels are more and more numerous, making it difficult to target candidates on the web. Existing decision support system on e-recruitment in the literature does not identify the desired profile from a job offer (C1), the relevance of a resume (C2) or the changing environment of recruitment (C3). Thereby, the objective of our research is to optimize the e-recruitment process by designing a decision support system capable of targeting potential candidates at a lower cost and that addresses the challenges (C1), (C2) and (C3).

Keywords: E-recruitment · Parsing · Matching · Machine learning · Decision support system

1 Introduction

The diversity of channels that broadcast job offers has expanded with the digital revolution (social networks, job boards, advertising sites). Each channel has a specific financial strategy and aims at targeting specific candidate profiles. As a result, online recruitment is becoming increasingly difficult for the recruiter. It is crucial to know precisely the characteristics of each channel. The broadcasting of job offers has a financing cost. For a company, the annual cost of recruitment can be very high. Consequently, it has become essential for recruiters to evaluate and analyze the various channels performances. The recruiter generally measures the performance of a recruitment campaign according to his objectives. Therefore, online recruitment requires analyzing massive data from multiple sources

© Springer Nature Switzerland AG 2021
I. Saad et al. (Eds.): ICIKS 2021, LNBIP 425, pp. 43–51, 2021.
https://doi.org/10.1007/978-3-030-85977-0_4

(career sites, recruitment sites, social network, etc.). As a result of these difficulties, some works in the literature have addressed e-recruitment optimization by proposing a recommendation system based on the content of the job offer [15] to estimate the performance of the channels. This work uses e-recruitment data containing interactions between the job offer and the candidates through the channels. These interactions generate events of clicks, applications sent etc. In this work, the conversion rate (the ratio between the number of resume received and the number of clicks) is used as an indicator, and the temporal dimension is not considered. However, it is evident that the moment when an offer is broadcasted on a channel influences its impact. The work of [16] has proved that taking temporality into account improves performance. Nevertheless, this system recommends channels based on the prediction of clicks over time. We have identified several weaknesses in the literature: (C1) The absence of the identification of the desired profile in job offers that would enable both to target more efficiently the most suitable candidates. A job offer can contain useless information that can create biases and ambiguities. Whereas the desired profile may be identified through the education, experience, skills etc. (C2) The exploitation of the numbers of clicks and the candidates' conversion rate. Recruiters use other indicators to define the performance of a recruitment campaign. For example, the cost per relevant candidate represents the cost of obtaining a resume that meets the desired profile in the advertisement. Although recruiters use this indicator, it is not taken into consideration in the literature. (C3) Several external factors (labor market, specific sector, etc.) or internal factors (more regular broadcasting on specific channels, human bias on the choice of channels) can influence clicks or the relevance of the resume received. These factors reflect a changing and uncertain environment. The uncertain nature of the environment is not considered in the literature. This research has one main objective: (1) Propose a decision support system that can (1) recommend the recruiter the channels that meet his objectives (2) address the limitations listed above. The proposed model should address the research question "How to optimize the selection of recruitment channels based on the desired profile and the resume relevance in a changing and uncertain environment?". The paper is organized as follows. Section 2 explore the machine learning approaches used in the literature for job offer parsing (C1), resume and job matching (C2) and finally the machine learning algorithm that can consider unknown external factors and the changing environment (C3). Section 3 defines the proposed approach. Section 4 presents the research perspectives and the discussion.

2   Related Work

2.1   Job Offer Parsing

The parsing aim is to identify and extract the desired profile from the offer by assigning the appropriate labels to the corresponding plain text: contract type, hard skills, required experience, etc. Several approaches have been studied in the literature to address this issue. **The rule-based approach** has been reviewed by [3] by combining regular expressions and dictionaries. Regular expressions that aim to identify strings, constructed with characters or meta-characters, were used to identify the postal code and the desired years of experience. The desired skills were identified using a dictionary containing the skills lexicon. This approach does indeed allow the extraction of years of experience and skills. However, its main limitation is that it cannot consider changes in the vocabulary (if a word is not represented in the dictionary of words or used in a regular expression, it cannot be parsed). Given the limitations of the rule-based, we turned to the machine learning approach. This approach is used in the process of knowledge extraction from data, and its objective is to extract information and exploit massive data. [8] propose an SVM-based **classification**. They use a vector representation for each text segment to assign a label to it. The SVM results do not provide the desired results due to various parameters (text written in natural language, uncertain splitting, varied delimiter). As a result, the analysis of the structure of the offer has made it possible to improve these results by considering the structure of the job offer. Nevertheless, we note two limitations to this work. The first is the choice of labels to be extracted (job, skills, salary, etc.) which is too limited to identify the desired profile. The second limitation is related to the use of the SVM, which does not exploit the plain text structure from the job offer. However, considering the structure could improve the parsing. The intuition behind our analysis is that recurrent neural networks can address this limitation as they are commonly used for plain text parsing in different domains. Due to the challenge of capturing dependencies between sequences in the long term, some works have favored LSTM (Long Short-Term Memory) recurrent neural networks that address the challenge of dependency between sequences. This approach can be applied to any textual document: natural language, structured, unstructured and semi-structured. **Sequence labeling** parsing considers a text as a sequence of sequences. The aim is to assign a label to each sequence. Our work's aim is to prove that this machine learning approach using sequence labeling associated with recurrent neural network can address the limitations of other approaches.

2.2   Job and Resume Matching

Numerous studies calculate the similarity between resume and job offers. The literature has examined rule-based methods, artificial learning, and knowledge management using ontologies. [5] proposes a **fuzzy** model for evaluating and selecting candidates based on skill. These skills are compared and ranked in comparison to the organization's objective. Only specific data (such as knowledge and skills) are used to sort the resume in [11]. All approaches tend to increase the effectiveness of recruiting applicants for jobs offers. Nonetheless, the full plain text of the job and resume is used. The plain texts, on the other hand, produce a significant amount of noise, resulting in poor precision and unsatisfactory rating performance. Certain approaches have favored an **ontology-based approach**. For instance, [17] proposes a method for automatically mapping a job offer ontology to a resume ontology. The rating is based on a similarity function between the two ontologies properties. Although this approach overcomes the limitations of manual matching of candidates, creating an ontology requires the processing of a large number of documents. Additionally, manual work by an expert is used to cover new skills or careers. According to [12] the matching problem is a component of a supervised learning framework in which **Deep Neural Networks** are used to identify the most qualified applicants for a job offer. The writers suggest that a convolutional neural network be adapted in a Siamese manner. The authors' method is based on a pairwise annotated dataset (label 0 if the job offer matches the resume, and 1 otherwise). Although the findings indicate that this technique performs well, we note a drawback with the annotated text pairs that need a considerable volume of annotated data to repeat these experiments. To address the annotation problem, [1] have developed a framework focused on deep learning that ranks job applicants according to their suitability for the job description. They accomplished this by using the BERT language representation model [4]. BERT was used to identify text segments and estimate the ranking of applicants for a work offer based on a similarity rate. Machine learning has significantly improved the matching of jobs and resumes. Present methods, on the other hand, are constrained by their syntactic structures and certain factors make reproducing the proposed approaches difficult. Recent study, which favors deep learning models, highlights the opacity of such systems. However, it is important to emphasize the recruiter's transparency about the matching system.

2.3   Machine Learning Algorithm Applied on Uncertain Environment

Reinforcement learning (RL) is a machine learning technique that allows an agent to learn by trial and error in an immersive environment using input from its own behaviors and experiences. Though both supervised and reinforcement learning use mapping between input and output, reinforcement learning utilizes incentives and punishment as cues for good and negative behaviour, unlike supervised learning. Reinforcement learning differs from unsupervised learning in terms of objectives. The aim of unsupervised learning is to identify parallels and discrepancies between data points. Whereas, the goal of reinforcement learning is to find an appropriate behavior model that maximizes the agent's overall accumulated reward. The fundamental theory and components of a reinforcement learning paradigm are shown in the diagram below. The following steps can characterise reinforcement learning: (1) The agent observes an input state and choose an action; (2) The action is performed; (3) The agent receives an outcome based on its environment; (4) Information about the given result for this state or action is recorded and the agent chooses a new action based on the reward of past actions. This approach has never been used in the field of recruiting. Additionally, it has been studied in similar areas such as film recommendation or foreign and investment operation. To refine suggestions in uncertain environments, reinforcement learning was used. As a result, a comparison to the recruiting domain is possible. The state of this system, on the other hand, should correspond to our performance indicators (clicks, number of relevant resume, and cost), the action should be the selection of one or more channels, and the reward should be the function that maximizes the number of relevant resume while minimizing clicks and cost. However, it is worthwhile to investigate this method further because it does not include historical data. This first benefit removes the restriction on the use of flawed statistics. Unlike reinforcement learning, which generates data at each learning iteration, **supervised learning** is a machine learning task that involves learning a prediction function from existing annotated dataset [13]. As a result, it requires a standardized, homogeneous, and annotated data corpus. In the case of recruiting, supervised learning has never been used. It will, however, allow the forecasting of performance indicators for a new job offer. These indicators' predictions will then be used to suggest channels that would increase the number of applications while reducing the number of clicks and cost. Supervised learning will overcome the first weakness identified in the state-of-the-art application by using these data as performance indicators.Furthermore, since this approach only considers historical data, it still discriminates against those channels, leaving out the factors that affect channel efficiency.

3   Proposed Approach

3.1   The Actors

The recruiter provides a job offer that includes a description of the desired profile. Different types of information can be found in the job offer to define the desired profile: profession, experience, education, hard skills, soft skills, missions, city, country, postal code, and contract type. It is important to define these labels. They are, in reality, critical for the job, resume parsing, and the matching of these two documents. The recruiter's company, its worth, and its size also reflect him. The recruiter also has a budget for e-recruitment, which we may refer to as financial constraints. **The channel** is represented by a type. It might be a social network like Facebook, Instagram, or a job board like Indeed, Glassdoor, or an advertising like Xander. The profile of users define also the characteristics of a channel. Indeed, different types of job boards may be used, each focusing on a particular contract type, number of years of experience, etc. **The candidate** is represented by his resume, which contains a huge amount of information. The same labels as the offers can be used to classify this information: occupation, experience, education, hard skills, soft skills, missions, city, country, and postal code.

3.2   Interaction Between Actors

The following describes the interactions between the various actors: (1) The recruiter sends job offers in a variety of formats to the system (xml, text, pdf etc.). Each offer or collection of offers is tied to a specific budget constraint. This data is saved in a database; (2) The decision support system (DSS) extracts data from the database and applies machine learning models to determine the optimal channel for each new job offer; (3) The recruiter is notified of the channel's ranking provided by the models; (4) The recruiter selects the channels and broadcast the job to the selected ones; (5) Once the jobs are posted on the channels, applicants can click to apply, submit their resume, or only display the job offer. (6) The tracker maintains a record of these events; (7) The DSM updates the learning models to reflect new events in order to update the channel recommendation to the recruiter.

3.3   Decision Support System

The proposed decision support system is composed of five components: **1. Data storage**. The aim of this component is to retrieve and store data from various channels. It's made up of a tracker that pulls every event from the job board. When an applicant, for example, clicks on a job offer on Indeed at time t, a new event is generated in the database. **2. Data processing**. Various types of data are processed in this component (Job offers and resume). The processing entails parsing the job offer and resumes to construct a standardized format for two purposes: (1) convert the plain text of the job offer into an xml format (the channels only support this data structure), (2) do the job and resume matching. Two of our challenges (C1) and (C2) are addressed by this component: **The parsing**: For its ability to view a set of words as a sequence, and hence the semantic meaning of words, sequence labeling has long been of special interest for natural language processing, such as part-of-speech tagging or semantic annotation, etc. [7,14]. The sequence labeling task entails assigning a categorical mark to each sequence using an algorithm that considers a text as a series of semantic words [9]. The parsed plain text is represented by the label/sequence pairs. The hypothesis of our work is that sequence labeling is an approach that can take advantage of plain text semi-structure while also considering the vocabulary evolution and ambiguity. The structure of the plain text (even if it isn't fixed) represents a collection of sequences and terms that can be used to capture meaning and thus consider vocabulary evolution. Our intuition is that sequence labeling could improve the parsing approaches used today to parse the job offers. **The matching**: Recent research on the matching has shown that machine learning methods can achieve a high level of accuracy. However, the recruiter cannot understand the results and implications of machine learning algorithms such as classifications or supervised learning. As a result, our primary research will concentrate on a hybrid system that will first construct a semantic vectorization using BERT in order to understand the word's semantic context. This vectorization will be applied to the parsing-generated labels/sequences. Second, we can compute a distance between the vectorized job offer labels/sequences and the resume. The similarity of two vector representations is determined by the distance between them. As a measure of similarity, we can use the cosine (R Baeza Yates, 1999), the Euclidean distance [6], or the Dice index [2]. These various techniques will be evaluated and only one will be kept. As a result, each label/sequence will have a distance score associated with it. This score would be more significant if the recruiter prioritizes this type of information over another. **3.Supervised learning**: this component uses the output of the job parsing, the events stored in the database to predict the future actions in each channel for each job offer. A deep neural network is used to predict these values. **4. Filtering**: is multi-objective function that aim to consider the objective of the recruiter. The first scenario is a single-objective optimization problem where the objective function is the maximization of the number of relevant resumes while adhering to aa cost constraint on each channel not to be exceeded. We then introduced an exponential smoothing technique, which is frequently used in time series research [10] to consider past data. The implementation of historical data enables the following: (1) The model's convergence; (2) Prevent the removal of a diffusion channel whose output measures are suboptimal for the time span T and are affected by an unknown factor (vacations, economic crisis, changing labor market etc.). The theory of exponential smoothing is to give greater weight to a data set's most recent observations. **5. Reinforcement learning**: Recall that reinforcement learning is characterized by the following steps defined in the state of the art the state of the art, steps that we instantiate on our problem: (1) The agent observes an input state. In our context the DSS is the agent that observes the new events coming from the channels at $t + T$; (2) The reward represents the number of relevant resume received in each channel; (3) An action is determined by the agent. The action in our model represents keep or change the channel; (4) The policy is the probability of taking action a for the next state for the context of the desired profile and the actual reward; (5) The action is performed; (6)The proposed action for each channel/job offer is displayed to the recruiter. This reinforcement learning module is re-launched periodically to re-evaluate the channels This reinforcement learning module is re-launched periodically to re-evaluate the channels chosen at initialization (we initially choose a period $T = 7$ days, but the value of T can obviously be set to another value) to assess the relevance of the channels and choose new actions accordingly. Thus, every day, the performance indicators are retrieved to infer their value over the period T. The reiteration of the choice of channels by learning by reinforcement at each period $t + T$ thus makes it possible to refine the results as the recruitment campaign progresses by learning from its actions/rewards and to adapt to the uncertain environment of the recruitment application context.

4   Research Perspectives

In conclusion, the state-of-the-art on e-recruitment optimization allowed us to define the limits of existing work in the literature, which we defined ahead of time through challenges C1, C2 and C3. We improved the state-of-the-art work by proposing a parsing model for the desired profile identification and the candidate profile from respectively the job offer and resume (C1). Following that, we used this parsing to create a semantically-aware matching between offers and resumes that allows us to explain the results to the recruiter using a distance measure between sequences and labels (C2). Following that, we focused our research on the design of a decision support system, proposing a hybrid system that relies on an initialization based on supervised learning in order to provide a fixed first model of recommendation that uses historical events. Given the limitations of supervised learning, we've added reinforcement learning to help with adaptation to an unknown environment and to avoid discrimination in some channels and human bias in the data (C3). We've defined our first function goal, which is critical for determining the pertinence of channels. This role will become more refined as more experiments are carried out. In the next part of our work, the objective is to use data from Xtramile, a digital recruitment company, in order to validate our models and our hypotheses.

References

1. Bhatia, V., Rawat, P., Kumar, A., Shah, R.R.: End-to-end resume parsing and finding candidates for a job description using BERT (2019)
2. Cabrera-Diego, L.A., Durette, B., Torres-Moreno, J.M., El-Bèze, M.: How can we measure the similarity between résumés of selected candidates for a job? July 2015
3. Casagrande, A., Gotti, F., Lapalme, G.: Cerebra, un système de recommandation de candidats pour l'e-recrutement. In: AISR2017, Paris, France, May 2017
4. Devlin, J., Chang, M.W., Lee, K., Toutanova, K.: BERT: pre-training of deep bidirectional transformers for language understanding (2019)
5. Golec, A., Kahya, E.: A fuzzy model for competency-based employee evaluation and selection. Comput. Ind. Eng. **52**(1), 143–161 (2007)
6. Gower, J.: Properties of Euclidean and Non-Euclidean distance matrices. Linear Algebra Appl. **67**, 81–97 (1985)
7. Kato, T., Abe, K., Ouchi, H., Miyawaki, S., Suzuki, J., Inui, K.: Embeddings of label components for sequence labeling: a case study of fine-grained named entity recognition. In: Proceedings of the 58th Annual Meeting of the Association for Computational Linguistics: Student Research Workshop, pp. 222–229. Online, July 2020
8. Kesler, R., Torres-Moreno, J.M., El-Bèze, M.: E-gen: traitement automatique d'informations de ressources humaines. Document numérique **13**, 95–119 (2010)
9. Lin, J.C.W., Shao, Y., Zhang, J., Yun, U.: Enhanced sequence labeling based on latent variable conditional random fields. Neurocomputing **403**, 431–440 (2020)
10. Liu, J., Si, Y.W., Zhang, D., Zhou, L.: Trend following in financial time series with multi-objective optimization. Appl. Soft Comput. **66**, 149–167 (2018)
11. Maheshwari, S., Sainani, A., Reddy, P.K.: An approach to extract special skills to improve the performance of resume selection. In: Kikuchi, S., Sachdeva, S., Bhalla, S. (eds.) DNIS 2010. LNCS, vol. 5999, pp. 256–273. Springer, Heidelberg (2010). https://doi.org/10.1007/978-3-642-12038-1_17
12. Maheshwary, S., Misra, H.: Matching resumes to jobs via deep Siamese network. In: Companion Proceedings of the The Web Conference 2018, WWW 2018, pp. 87–88. International World Wide Web Conferences Steering Committee, Republic and Canton of Geneva, CHE (2018)
13. Nasteski, V.: An overview of the supervised machine learning methods. HORIZONS.B **4**, 51–62 (2017)

14. Ramponi, A., van der Goot, R., Lombardo, R., Plank, B.: Biomedical event extraction as sequence labeling. In: Proceedings of the 2020 Conference on Empirical Methods in Natural Language Processing (EMNLP), pp. 5357–5367 (2020)
15. Séguéla, J., Saporta, G.: A hybrid recommender system to predict online job offer performance. Revue des Nouvelles Technologies de l'Information **RNTI -E-25**, 177–197 (2013). Special issue: HSDA 2013, Advances in Theory and Applications of High Dimensional and Symbolic Data Analysis
16. Sidahmed, B., Mellouli, N., Lamolle, M.: On the predictive analysis of behavioral massive job data using embedded clustering and deep recurrent neural networks. Knowl.-Based Syst. **151**, 95–113 (2018)
17. Senthil Kumaran, A., Sankar, A.: Towards an automated system for intelligent screening of candidates for recruitment using ontology mapping expert. Int. J. Metadata Semant. Ontol. **8**, 56–64 (2013)

Cognitive Effort Reduction Within Group Decision Making Through Aggregation and Disaggregation of Individual Preferences

Salem Chakhar[1,2(✉)], Inès Saad[3,4], Ashraf Labib[1,2], and Alessio Ishizaka[5]

[1] Portsmouth Business School, University of Portsmouth, Portsmouth, UK
[2] Centre for Operational Research and Logistics, University of Portsmouth, Portsmouth, UK
`{salem.chakhar,ashraf.labib}@port.ac.uk`
[3] MIS Laboratory, University of Picardie Jules Verne, Amiens, France
[4] ESC Amiens, Amiens, France
`ines.saad@esc-amiens.com`
[5] NEOMA Business School, Mont-Saint-Aignan, France
`alessio.ishizaka@neoma-bs.fr`

Abstract. Most of multicriteria methods require the specification of different preference parameters. This necessitates a considerable cognitive effort from the decision maker, especially for those with no or limited knowledge about multicriteria analysis. The situation is further complicated within a group of decision makers. In this paper, we propose a multicriteria classification approach that relies on an aggregation/disaggregation strategy within a group of decision makers, permitting thus to considerably reduce the needed cognitive effort at individual as well as group levels. The aggregation/disaggregation strategy is known for its ability to reduce the cognitive effort at the level of individual decision makers. At group level, a simple majority rule is used to generate consensual decisions with no additional information from the decision makers. A didactic example is used to illustrate the proposed approach.

Keywords: Group decision making · Aggregation/Disaggregation approach · Cognitive effort · Majority principle

1 Introduction

A relatively high number of multicriteria methods are currently available in the literature [14]. Most of these methods require the specification of several preference parameters such as criteria weights and discrimination thresholds. However, it is known that a decision maker often has difficulties to specify precise values for preference parameters and feel more comfortable to give holistic judgements [22]. These difficulties are mainly due to the cognitive load and mental resources [17] needed to explicitly specify values of these parameters. Such mental exercise

© Springer Nature Switzerland AG 2021
I. Saad et al. (Eds.): ICIKS 2021, LNBIP 425, pp. 52–67, 2021.
https://doi.org/10.1007/978-3-030-85977-0_5

clearly requires an important cognitive effort from the decision maker, particularly from those with no or limited knowledge about multicriteria analysis. The situation is further complicated within a group of decision makers where the specification of these parameters requires a high level of agreement between involved decision makers and faces several conflicting situations [4].

One possible solution to reduce the cognitive effort is to use an aggregation/disaggregation approach to infer values of these parameters [11,12]. The basic input for the aggregation/disaggregation approach is a set of assignment examples specified by the decision maker. Then, an inference procedure is used to deduce, through a certain form of regression, a set of preference parameters values that re-do at best the assignment examples. Aggregation/disaggregation approach has been used largely for single decision makers [1,7,10,16,18]. There are also some group oriented aggregation/disaggregation approaches [2,6,15]. For instance, the study reported in [6] proposes a methodology in which a group of decision makers discuss how to sort some assignment examples, instead of discussing what values the preference parameters should take, which needs a high level of agreement between the decision makers. Generally, most of existing proposals proceed either with input or output aggregation strategy that are both characterized by some shortcomings, as discussed in [4,5]. Furthermore, most of group oriented aggregation/disaggregation approaches need a strong agreement between decision makers and often proceed iteratively to reach a consensual decision.

The objective of this paper is to propose an aggregation/disaggregation-based approach to support group decision making in multicriteria classification problems. A multicriteria classification problem is sharply defined as the assignment of a set of decision objects described with respect to several evaluation criteria into a set of predefined and preference-ordered decision classes. The proposed approach is organized into four phases: (1) individual classification, (2) aggregation, (3) conflict resolving, and (4) collective classification. In the first phase, each decision maker runs individually the inference procedure using as input his/her own assignment examples. In the second phase, an assignment rule—which is based on majority principle—is used to construct a set of collective assignment examples. The third phase uses some coherence rules to resolve conflicting situations. The forth phase applies the inference procedure using the collective assignment examples as input in order to infer values for collective preference parameters, which are then used to assign decision objects to decision classes. A didactic example is used to illustrate the proposed approach.

The paper goes as follows. Section 2 presents the background. Section 3 introduces a general outline of proposed approach. Section 4 details the aggregation procedure. Section 5 addresses the problem of conflict resolving. Section 6 provides a didactic example. Section 7 concludes the paper.

2 Background

2.1 Basic Notations

Let $U = \{x_k : k = 1, \ldots, n\}$ be a finite set of n decision objects and $Q = \{q_j : j = 1, \ldots, m\}$ be a set of m evaluation criteria. The evaluation of an object $x \in U$ according to criterion $q_j \in Q$ is written $q_j(x)$. Without loss of generality, we assume that the preference is increasing with value of $q_j(.)$ for every j. Let $Cl = \{Cl_t : t = 1, \ldots, p\}$ be a set of p decision classes. We suppose that the classes are preference-ordered, i.e., for all $Cl_r, Cl_s \in Cl$, such that $r > s$, the objects from Cl_r are preferred to the objects from Cl_s.

The decision classes are defined through a set of $p - 1$ limiting profiles $L = \{b_t : t = 1, \ldots, p - 1\}$, where b_t is the upper limit of class Cl_t and lower limit of class Cl_{t+1}.

The assignment of decision objects to decision classes relies on a multicriteria classification method. Multicriteria (classification) methods often require the definition of a set of preference parameters such as criteria weights and thresholds. In this paper, a multicriteria classification method is generically designed by $\Gamma_w : U \times Q \rightarrow Cl$ where Γ is a multicriteria classification method and w is a set of preference parameters required to apply Γ.

2.2 Principles of Aggregation/Disaggregation Approach

The general schema of an aggregation/disaggregation approach can be structured into two stages: (i) inference of preference parameters values, and (ii) exploitation and classification. The objective of the first stage is to infer the values of a set w of preference parameters using as input some holistic information provided by the decision maker. The holistic information are global judgements on decision objects and thus represent *aggregated* information. The inferred parameters values are obtained by *disaggregating* the global information provided by the decision maker. At the end of the first stage, the decision maker should agree on the inferred values. Otherwise, the first stage can be restarted again by using different input data. The objective of the second stage is to use a multicriteria classification method Γ_w (the values of preference parameters set w are those obtained in the first stage) to assign the decision objects to different classes.

The holistic information may take different forms such as assigning objects to decision classes, constraints on preference parameters, etc. In this paper, holistic information are represented as a set of assignment examples $U^* \in U$ defined such that each $x \in U^*$ is assigned to a range $[Cl_{t_1}, Cl_{t_2}]$ of possible classes where Cl_{t_1} and Cl_{t_2} represent respectively the minimum and maximum classes to which x should be assigned. The assignment of x to a precise class Cl_t can be modeled by setting $Cl_{t_1} = Cl_{t_2} = Cl_t$.

2.3 Inference Procedure

The main component of the aggregation/disaggregation approach is the inference procedure, denoted by $\mathcal{I}$ in the rest of this paper, which is used to induce the values for the preference parameters set w. The basic idea of the inference procedure consists in finding a set of preference parameters values that permit to re-do the assignment examples provided by the decision maker. The inference procedure proceeds as follows. First, let S be an outranking relation defined such that xSb_t means that the evaluations of x upon all criteria in Q are at least as good as the evaluations of the limiting profile b_t (lower limit of class Cl_{t+1}). Based on the assignment examples set U^*, we then define two sets as follows:

- $S^+ = \{(x, b_t) \in U^* \times L$ and the decision maker states that $uSb_t\}$.
- $S^- = \{(u, b_t) \in U^* \times L$ and the decision maker states that $\neg(uSb_t)\}$.

This information are then used to define a mathematical program such that: (i) the decision variables are the preference parameters to infer, (ii) the objective function is defined as an error function measuring the level to which the assignment examples are restituted, and (iii) the constraints express restrictions on the values of preference parameters. The values for preference parameters are then obtained by maximizing the minimum slack for this system of constraints.

In addition to specifying assignment examples, the decision maker can also fix the values of some preference parameters. The ability to fix the values of some preference parameters is very useful in practice since jointly inferring all preference parameters as in [19]) often requires solving a non-linear mathematical program, which is time demanding. To obtain a linear mathematical programs, it is possible to use partial inference procedures where only a subset of preference parameters are inferred as criteria weights and the cutting level [10], veto thresholds [9] and category limits [20].

3   Approach Outline

The proposed approach is composed of four phases: (1) individual classification, (2) aggregation, (3) conflict resolving, and (3) collective classification. A brief description of these phases follows. Let $H = \{d_i : i = 1, \ldots, h\}$ a finite set of h decision makers.

3.1   Individual Classification

In this phase, each decision maker d_i applies the inference procedure using as input his/her own assignment examples. It is assumed that decision makers are using the same inference procedure $\mathcal{I}$. Further, we assume that all decision makers agree on the values of fixed preference parameters and the constraints these parameters. The assignment examples of each decision maker d_i can be represented through as individual assignment matrix $A_i[n \times p]$. Each element of the assignment matrix A_i is defined for each $x_k \in U$ and $Cl_t \in Cl$ as follows:

$$A_i(x_k, Cl_t) = \begin{cases} 1, \text{ if } x_k \text{ is assigned by the decision maker } d_i \text{ to } Cl_t, \\ 0, \text{ otherwise.} \end{cases} \quad (1)$$

The output of the application of the inference procedure on matrix A_i can be represented through an individual classification matrix $R_i[n \times p]$. Each element of the individual classification R_i is defined for each $x_k \in U$ and $Cl_t \in Cl$ as follows:

$$R_i(x_k, Cl_t) = \begin{cases} 1, \text{if } x_k \text{ is assigned by the inference procedure } \mathcal{I} \text{ to } Cl_t, \\ 0, \text{otherwise.} \end{cases} \quad (2)$$

The inference procedure $\mathcal{I}$ may assign an object $x_k \in U$ to a range of possible assignments of the form $[W(x_k), B(x_k)]$ where $W(x_k)$ and $B(x_k)$ indicate the worst and best classes to which x_k can be assigned, receptively. When the inputs are consistent, the inference procedure $\mathcal{I}$ is able to identify for each object $x_k \in U$ 'central' assignment $C(x_k) \in [W(x_k), B(x_k)]$ that minimizes the maximum slack. When the inputs are inconsistent, only a single assignment is computed by inference procedure $\mathcal{I}$. Based on the output of the inference procedure $\mathcal{I}$, we construct an individual central classification matrix $R_i^*[n \times p]$. Each element of the individual central classification R_i^* is defined for each $x_k \in U$ and $Cl_t \in Cl$ as follows:

$$R_i^*(x_k, Cl_t) = \begin{cases} 1, \text{if } R_i(x_k, Cl_t) = 1 \text{ and } Cl_t = C(x_k), \\ 0, \text{otherwise.} \end{cases} \quad (3)$$

3.2   Aggregation

The objective of this phase is to construct a collective assignment matrix A by aggregating individual assignment matrices $A_1, A_2, \cdots, A_h$. The elements of the collective assignment matrix A are defined for each $x_k \in U$ and $Cl_t \in Cl$ as follows:

$$A(x_k, Cl_t) = \begin{cases} 1, & \text{if a majority of decision makers support } x_k \in Cl_t, \\ 0, & \text{if a majority of decision makers reject } x_k \in Cl_t, \\ \text{`?'}, & \text{otherwise.} \end{cases} \quad (4)$$

The two first cases are straightforward. The third case corresponds to the situation where there is not any majority neither in favor of the assignment of x_k to Cl_t nor in favor of the non-assignment of x_k to Cl_t. We denote this conflicting situation by '?'. The conflicting situations will be treated in the fourth phase.

The construction of the collective assignment matrix A uses an aggregation procedure that will be detailed in Sect. 4. This procedure relies on an assignment rule that coherently implement the majority principle. The contribution of each decision maker in the collective decision is objectively measured by the quality of individual classification conducted by this decision maker.

3.3   Conflict Resolving

The objective of this phase is to resolve the conflicting situations in the collective assignment matrix A. In this paper, we designed some simple coherence rules than can be used to automatically fix the conflicting situations. In terms of this phase, we obtain a revised collective assignment matrix A' defined as follows:

$$A'(x_k, Cl_t) = \begin{cases} 1, \text{ if a majority of decision makers support } x_k \in Cl_t, \\ 0, \text{ otherwise.} \end{cases} \quad (5)$$

The problem of conflict resolving is detailed in Sect. 5.

3.4   Collective Classification

The objective of this phase is to apply the inference procedure $\mathcal{I}$ using the revised collective assignment matrix A' as input. The application of $\mathcal{I}$ at this level is the same as with a single decision maker. The main output of this phase is the values for non-fixed preference parameters w. The induced preference parameters values are then used to run the multicriteria classification model Γ_w to obtain the final classification.

4   Aggregation Procedure

The aggregation procedure relies on majority principle, which is defined by means of concordance powers and implemented through an assignment rule. The definition of concordance power requires the introduction of a new metric P that characterizes the information provided by each decision maker.

4.1   Definition of Metric P

We define first matrix $M_i[n \times p]$ $(i = 1, \cdots, h)$ as follows:

$$M_i(x_k, Cl_t) = \begin{cases} 0, & \text{if } x_k \notin U_i^* \vee [(x_k \in U_i^* \wedge A_i(x_k, Cl_t) = R_i(x_k, Cl_t)) \wedge \\ & \quad R_i(x_k, Cl_t) = R_i^*(x_k, Cl_t)]. \\ \beta, & \text{if } x_k \in U_i^* \wedge A_i(x_k, Cl_t) = R_i(x_k, Cl_t)) \wedge \\ & \quad R_i(x_k, Cl_t) \neq R_i^*(x_k, Cl_t)]. \\ \alpha, & \text{if } x_k \in U_i^* \wedge A_i(x_k, Cl_t) = 0 \wedge R_i(x_k, Cl_t) = 1. \\ 1 - \alpha, & \text{if } x_k \in U_i^* \wedge A_i(x_k, Cl_t) = 1 \wedge R_i(x_k, C_t) = 0. \end{cases} \quad (6)$$

where $U_i^* \subseteq U$ is the set of assignment examples provided by decision maker d_i, and α are β penalty parameters defined such that: $0 < \alpha \leq 1$ and $0 \leq \beta < \min\{\alpha, 1 - \alpha\}$. The first case in the definition of M_i corresponds to exact assignments, i.e., the assignment provided by the decision maker d_i is the same as the central assignment generated by the inference procedure $\mathcal{I}$. The second case is as the previous one but the assignment provided by the decision maker is within the range $[W(x_k), B(x_k)]$ but different to the central one $C(x_k)$. In this case, a small penalty of β is applied. The third case represents a situation where decision maker d_i do not assign object x_k to class Cl_t while the inference procedure $\mathcal{I}$ does. Thus, α represents the penalty of a missing assignment. The fourth case represents the situation where decision maker d_i assigns object x_k to class Cl_t while the inference procedure $\mathcal{I}$ does not. Thus, $1 - \alpha$ represents the penalty of a wrong assignment.

The parameter α as defined above permits to penalize differently wrong assignments and missing assignments. Indeed, depending on the application domain, wrong and missing assignments may have different consequences. In domains where 'no decision' is better than 'wrong decision' as in human health, we should define α such that $\alpha < \frac{1}{2}$, which penalizes wrong assignments more than missing assignments. Defining α such that $\alpha > \frac{1}{2}$, gives more penalty to missing assignments than wrong assignments. Setting $\alpha = \frac{1}{2}$ gives the same penalty to missing and wrong assignments.

The level to which the assignment examples provided by decision maker d_i are reproduced by the inference procedure $\mathcal{I}$ can be measured through function $P : H \rightarrow \mathbf{R}^+$ defined as follows:

$$P(d_i) = \sum_{k=1}^{n} \sum_{t=1}^{p} M_i(x_k, Cl_t) \tag{7}$$

4.2   Concordance Power

First, we need to standardize $P(d_i)$ $(i = 1, \cdots, h)$ using the following formula:

$$P'(d_i) = \frac{P(d_i)}{\sum_{r=1}^{h} P(d_r)}, \quad (i = 1, \cdots, h). \tag{8}$$

We then define set F_a for all $a \in \{0, 1\}$ as follows:

$$F_a(x_k, Cl_t) = \{i : d_i \in H \wedge A_i(x_k, Cl_t) = a\} \tag{9}$$

The concordance power for assigning an object $x_k \in U$ to a class $Cl_t \in Cl$ can then be computed by:

$$\Pi_a(x_k, Cl_t) = \sum_{i \in H} \Pi_a^i(x_k, Cl_t) \tag{10}$$

with

$$\Pi_a^i(x_k, Cl_t) = \begin{cases} \frac{1 - P'(d_i)}{\sum_{r \in F_a(x_k, Cl_t)} 1 - P'(d_r)}, & \text{if } i \in F_a, \\ 0, & \text{otherwise.} \end{cases} \tag{11}$$

The number $\Pi_a(x_k, Cl_t)$ can be seen as the 'power' of the coalition of decision makers that support assertion $A_t(x_k, Cl_t) = a$. The concordance power $\Pi_1(x_k, Cl_t)$ can then be seen as an argument that *supports* the assignment of x_k to Cl_t, i.e., $A_t(x_k, Cl_t) = 1$, while concordance power $\Pi_0(x_k, Cl_t)$ can be seen as an argument that *rejects* the assignment of x_k to Cl_t, i.e., $A_t(x_k, Cl_t) = 0$.

4.3   Assignment Rule

The assignment rule is defined based on the concordance power $\Pi_a(x_k, Cl_t)$. Let $\theta \in [0.5, 1]$ be a majority threshold. Then, the assignment rule is define as follows:

Assignment Rule 1 $\quad \left| \begin{array}{l} \textbf{if} \quad \Pi_a(x_k, Cl_t) \geq \theta \textbf{ then } A(x_k, Cl_t) = a \\ \textbf{else } A(x_k, Cl_t) = '?' \end{array} \right.$

This assignment rule implements three possible situations:

- When there is a majority in favor of assigning x_k to Cl_t, then $\Pi_1(x_k, Cl_t) \geq \theta$ and, consequently, $\Pi_0(x_k, Cl_t) < 1 - \Pi_1(x_k, Cl_t)$. In this case, it is reasonable to $A(x_k, Cl_t) = 1$.
- When there is a majority against assigning x_k to Cl_t, then $\Pi_0(x_k, Cl_t) \geq \theta$ and, consequently, $\Pi_1(x_k, Cl_t) < 1 - \Pi_0(x_k, Cl_t)$. In this case, it is reasonable to $A(x_k, Cl_t) = 0$.
- When there is no majority in favor or against assigning x_k to Cl_t, then $\Pi_1(x_k, Cl_t) < \theta$ and $\Pi_0(x_k, Cl_t) < \theta$. In this case, the majority rule cannot be used and we set $A(x_k, Cl_t) = '?'$. Some coherence rules can then be used to specify the value of $A(x_k, Cl_t)$.

5   Conflict Resolving

A conflict situation holds if there is no majority in favor or against the assignment of decision object x_k to decision class Cl_t. This situation has been designed by the symbol '?' in the initial collective assignment matrix. A trivial solution to conflict resolving is to call decision makers to modify their assignment examples in such that a majority in favor or against the assignment holds. This is a time-consuming exercise and requires a high cognitive effort from the decision makers. Thus, to reduce the cognitive effort required from them, we propose to use some simple coherence rules to automatically fix the conflicting situations.

The first conference rule is defined as follows:

Coherence Rule 1 $\quad \left| \begin{array}{l} \textbf{if} \quad A(x, Cl_1) = '?' \wedge \ldots \wedge A(x_k, Cl_p) = '?' \\ \textbf{then } A'(x, Cl_1) = 1 \wedge \ldots \wedge A'(x_k, Cl_p) = 1 \end{array} \right.$

This coherence rule says that if no majority can be established for assigning decision object x_k to any decision classes, then decision object x_k is removed from the assignment examples set U^*. In this case, decision object x_k is assigned by default to all decision classes.

The second and third coherence rules are defined as follows for $a \in \{0, 1\}$:

Coherence Rule 2 $\left|\begin{array}{l}\textbf{if} \quad A(x, Cl_{t \geq t'}) = a \wedge A(x_k, Cl_{t < t'}) = \text{`?'} \\ \textbf{then } A'(x_k, Cl_{t \geq t'}) = a \wedge A'(x_k, Cl_{t < t'}) = \neg a\end{array}\right.$

Coherence Rule 3 $\left|\begin{array}{l}\textbf{if} \quad A(x, Cl_{t \leq t'}) = a \wedge A(x_k, Cl_{t > t'}) = \text{`?'} \\ \textbf{then } A'(x_k, Cl_{t \leq t'}) = a \wedge A'(x_k, Cl_{t > t'}) = \neg a\end{array}\right.$

The second and third coherence rules apply when a majority holds for a range of decision classes and do not hold for the other range of decision classes.

A forth coherence rule is defined as follows:

Coherence Rule 4 $\left|\begin{array}{l}\textbf{if} \quad A(x, Cl_{t_1 \leq t \leq t_2}) = a) \wedge \\ \quad (A(x_k, Cl_{t \notin [t_1, t_2]}) = \neg a \vee A(x_k, Cl_{t \notin [t_1, t_2]}) = \text{`?'}) \\ \textbf{then } A(x_k, Cl_t) = a\end{array}\right.$

This forth coherence rule can be seen as a generalization of the second and third coherence rules.

If none of these coherence rules applies for a given decision object, then we can ask the decision makers to modify the corresponding assignment example.

6   Illustrative Didactic Example

This section illustrates the proposed approach through a didactic example. In this application, the Interactive Robustness analysis and Inference for Sorting problems (IRIS) tool [8] has been used to infer the preferences parameters and conduct individual classifications. IRIS supports an aggregation-disaggregation approach tailored around a pessimistic version of ELECTRE TRI method [13].

6.1   Decision Problem and Dataset

We consider the problem of assigning ten students (S1 to S10) into three preference ordered decision classes. The datatset in Table 1 represents the assessment of these students with respect to four criteria and their assignments by three decision makers (namely DM1, DM2 and DM3) into three decision classes. Each student has been assigned by the decision makers into a range $[l_i, u_i]$ $(i = 1, 2, 3)$ where l_i and u_i are the lower and higher class where he/she may be assigned according to decision maker d_i. Typically, each student is assigned by default to the range [1,3]. If the decision maker changes these values, then the student's assignment will be constrained and it becomes an assignment example.

The limiting profiles are given in Table 2 along with the indifference and preference thresholds over all criteria. The veto threshold will not be used. The criteria weights and cutting level parameters are unknown and should be inferred. We finally assume that ELECTRE TRI is used as the multicriteria classification method.

Table 1. Dataset.

Student	Mathematics	Physics	Literature	Philosophy	l_1	u_1	l_2	u_2	l_3	u_3
S1	3	2	1	2	2	2	1	1	2	2
S2	2	2	1	1	1	2	1	2	1	2
S3	2	2	2	1	1	3	1	3	1	2
S4	3	3	2	2	2	2	2	3	1	3
S5	3	3	2	2	1	3	1	3	2	3
S6	3	2	3	3	2	3	3	3	1	3
S7	3	3	3	2	1	3	1	3	2	3
S8	1	1	1	1	1	3	1	1	1	3
S9	1	1	2	1	1	3	1	1	1	2
S10	3	2	2	1	2	2	1	2	1	3

Table 2. Specification of profiles.

Profiles	Mathematics	Physics	Literature	Philosophy
Profile 1	2	2	2	2
Indifference	0	0	0	0
Preference	1	1	1	1
Profile 2	3	3	3	3
Indifference	0	0	0	0
Preference	1	1	1	1

6.2 Application and Results

Individual Classification. The assignment examples in Table 1 can be represented through initial assignment matrices as shown in Table 3. Each decision maker has then used his/her assignment examples as input to IRIS to infer the non-fixed preference parameters. The results of inference are summarized in Table 4. The inferred parameters values are then used by IRIS to re-assign the students into the decision classes. The results can then be represented via the individual classification matrices shown Table 5. The corresponding individual central classification matrices are given in Table 6.

Table 3. Individual assignment matrices.

A_1

Student	Cl_1	Cl_2	Cl_3
S1	0	1	0
S2	1	1	0
S3	1	1	1
S4	0	1	0
S5	1	1	1
S6	0	1	1
S7	1	1	1
S8	1	1	1
S9	1	1	1
S10	0	1	0

A_2

Student	Cl_1	Cl_2	Cl_3
S1	1	0	0
S2	1	1	0
S3	1	1	1
S4	0	1	1
S5	1	1	1
S6	0	0	1
S7	1	1	1
S8	1	0	0
S9	1	0	0
S10	1	1	0

A_3

Student	Cl_1	Cl_2	Cl_3
S1	0	1	0
S2	1	1	0
S3	1	1	0
S4	1	1	1
S5	0	1	1
S6	1	1	1
S7	0	1	1
S8	1	1	1
S9	1	1	0
S10	1	1	1

Table 4. Inferred values during individual classification.

Decision maker	Weights				Credibility threshold
	Mathematics	Physics	Literature	Philosophy	
DM1	0.375	0.125	0.250	0.250	0.625
DM2	0.333	0	0.333	0.334	0.833
DM3	0.300	0.300	0.100	0.300	0.700

Table 5. Individual classification matrices.

R_1

Student	Cl_1	Cl_2	Cl_3
S1	0	1	0
S2	1	0	0
S3	0	1	0
S4	0	1	0
S5	0	1	0
S6	0	1	1
S7	0	0	1
S8	1	0	0
S9	1	0	0
S10	0	1	0

R_2

Student	Cl_1	Cl_2	Cl_3
S1	1	0	0
S2	1	0	0
S3	1	1	0
S4	0	1	0
S5	0	1	0
S6	0	0	1
S7	0	1	1
S8	1	0	0
S9	1	0	0
S10	1	1	0

R_3

Student	Cl_1	Cl_2	Cl_3
S1	0	1	0
S2	1	1	0
S3	1	1	0
S4	0	1	1
S5	0	1	1
S6	0	1	1
S7	0	0	1
S8	1	0	0
S9	1	0	0
S10	1	1	0

Table 6. Individual central classification matrices.

R_1^*

Student	Cl_1	Cl_2	Cl_3
S1	0	1	0
S2	1	0	0
S3	0	1	0
S4	0	1	0
S5	0	1	0
S6	0	0	1
S7	0	0	1
S8	1	0	0
S9	1	0	0
S10	0	1	0

R_2^*

Student	Cl_1	Cl_2	Cl_3
S1	1	0	0
S2	1	0	0
S3	1	0	0
S4	0	1	0
S5	0	1	0
S6	0	0	1
S7	0	1	0
S8	1	0	0
S9	1	0	0
S10	1	0	0

R_3^*

Student	Cl_1	Cl_2	Cl_3
S1	0	1	0
S2	1	0	0
S3	0	1	0
S4	0	1	0
S5	0	1	0
S6	0	0	1
S7	0	0	1
S8	1	0	0
S9	1	0	0
S10	0	1	0

Aggregation. To compute metric $P(d_i)$ $(i = 1, 2, 3)$, we need first to construct matrices M_i $(i = 1, 2, 3)$. The results are given in Table 7 for $\alpha = 0.74$ and $\beta = 0.2$. The values for metric $P(d_i)$ $(i = 1, 2, 3)$ for the three decision makers are then computed and the obtained results are summarised in Table 8. This table also shows the values of $P'(d_i)$ $(i = 1, 2, 3)$.

The concordance powers are then computed and the corresponding results are shown in Table 9. The initial collective assignment matrix given in Table 10 is construed based on the concordance powers using a majority threshold of $\theta = 0.8$.

Table 7. Matrices M_i $(i = 1, 2, 3)$.

M_1 Student	Cl_1	Cl_2	Cl_3
S1	0	0	0
S2	0	0.26	0
S3	0	0	0
S4	0	0	0
S5	0	0	0
S6	0	0.2	0
S7	0	0	0
S8	0	0	0
S9	0	0	0
S10	0	0	0

M_2 Student	Cl_1	Cl_2	Cl_3
S1	0	0	0
S2	0	0.26	0
S3	0	0	0
S4	0	0	0.26
S5	0	0	0
S6	0	0	0
S7	0	0	0
S8	0	0	0
S9	0	0	0
S10	0	0.2	0

M_3 Student	Cl_1	Cl_2	Cl_3
S1	0	0	0
S2	0	0.2	0
S3	0.2	0	0
S4	0	0	0
S5	0	0	0.2
S6	0	0	0
S7	0	0.26	0
S8	0	0	0
S9	0	0.26	0
S10	0	0	0

Table 8. Values of $P(d_i)$ and $P'(d_i)$ $(i = 1, 2, 3)$.

	d_i		
	DM1	DM2	DM2
$P(d_i)$	0.46	0.72	1.12
$P'(d_i)$	0.2	0.313	0.487

Table 9. Concordance powers.

Π_0 Student	Cl_1	Cl_2	Cl_3
S1	0.657	0.343	1
S2	0	0	1
S3	0	0	0.257
S4	0.743	0	0.4
S5	0.257	0	0
S6	0.743	0.343	0
S7	0.257	0	0
S8	0	0.343	0.343
S9	0	0.343	0.6
S10	0.4	0	0.743

Π_1 Student	Cl_1	Cl_2	Cl_3
S1	0.343	0.657	0
S2	1	1	0
S3	1	1	0.743
S4	0.257	1	0.6
S5	0.743	1	1
S6	0.257	0.657	1
S7	0.743	1	1
S8	1	0.657	0.657
S9	1	0.657	0.4
S10	0.6	1	0.257

Table 10. Initial collective assignment matrix.

A Student	Cl_1	Cl_2	Cl_3
S1	?	?	0
S2	1	1	0
S3	1	1	?
S4	?	1	?
S5	?	1	1
S6	?	?	1
S7	?	1	1
S8	1	?	?
S9	1	?	?
S10	?	1	?

Conflict Resolving. Relying on the designed coherence rules, the conflicting situations in the initial collective assignment matrix (in Table 10) have been solved and the corresponding final collective assignment matrix is given in Table 11.

Table 11. Final collective assignment matrix.

A'			
Student	Cl_1	Cl_2	Cl_3
S1	1	1	0
S2	1	1	0
S3	1	1	0
S4	0	1	0
S5	0	1	1
S6	0	0	1
S7	0	1	1
S8	1	0	0
S9	1	0	0
S10	0	1	0

Collective Classification. Finally, the collective assignment matrix has been used as input to IRIS to infer the values of preference parameters at group level. The obtained values are given in Table 12. These values are then used to apply ELECTRE TRI. The obtained final assignments of students are given in Table 13.

Table 12. Inferred values during collective classification.

Weights				Credibility threshold
Mathematics	Physics	Literature	Philosophy	
0.333	0.167	0.333	0.167	0.667

Table 13. Final assignment results.

Student	Class		
	Worst	Best	Central
S1	1	2	2
S2	1	1	1
S3	2	2	2
S4	2	2	2
S5	2	2	2
S6	3	3	3
S7	3	3	3
S8	1	1	1
S9	1	1	1
S10	2	2	2

6.3 Evaluation

Table 14 presents the correlation analysis of individual and group central assignments. Individual central assignments are those obtained by different decision makers at the end of the individual classification phase while group central assignments are the final central assignments generated by the proposed approach. The figures in Table 14 indicate the proposed approach perfectly reproduced the individual central assignments of decision makers DM1 and DM3, and about 70% of the individual central assignments of decision maker DM2.

Table 14. Correlation analysis of individual and group central assignments

Kendall's τ	DM1	DM2	DM3	Group	Spearman's ρ	DM1	DM2	DM3	Group	Pearson's r	DM1	DM2	DM3	Group
DM1	1	0.691	1	1	DM1	1	0.730	1	1	DM1	1	0.745	1	1
DM2	0.691	1	0.691	0.691	DM2	0.730	1	0.730	0.730	DM2	0.745	1	0.745	0.745
DM3	1	0.691	1	1	DM3	1	0.730	1	1	DM3	1	0.745	1	1
Group	1	0.691	1	1	Group	1	0.730	1	1	Group	1	0.745	1	1

7 Conclusion

The paper introduced and illustrated a four phase aggregation/disaggregation approach for multicriteria classification for group decision making. The proposed approach uses a mixed input-output strategy, eliminating thus most of shortcomings encouraged by input or output aggregation strategies. In this sense, it follows the idea proposed in [3,5]. In addition, the proposed approach needs no collaboration and a very limited agreement between decision makers. This is because the final consensual decisions are automatically generated, with no intervention from the decision makers. This is very suitable in practice when there is a strong disagreement and/or high relational conflict between decision makers. Additionally, the construction of consensual decisions rely on simple majority rule, which—as advocated by [21]—satisfies anonymity and neutrality. The way the majority rule has been defined allows it to take into account the quality of assignments provided by each decision maker.

Several topics need to be further investigated. From practical point of view, we first intend to apply the proposed approach in real-life decision problems. Another topic concerns the application of the proposed approach with other multicriteria classification methods than ELECTRE TRI used in this paper. A third topic is related to the study of the computational behavior of the proposed approach with large datasets. Finally, we intend to enhance the assignment rule, by considering both the majority principle and veto effect.

References

1. Alvarez, P., Lopez, J., Parra, P.: A new disaggregation preference method for new products design. In: Liu, J., Lu, J., Xu, Y., Martinez, L., Kerre, E.E. (eds.) Data Science and Knowledge Engineering for Sensing Decision Support, pp. 1010–1017. World Scientific (2018)
2. Bregar, A.: Application of a hybrid Delphi and aggregation-disaggregation procedure for group decision-making. EURO J. Decis. Process. **7**, 3–32 (2019)
3. Chakhar, S., Saad, I.: Dominance-based rough set approach for groups in multicriteria classification problems. Decis. Support Syst. **54**(1), 372–380 (2012)
4. Chakhar, S., Saad, I.: Incorporating stakeholders' knowledge in group decision-making. J. Decis. Syst. **23**(1), 113–126 (2014)
5. Chakhar, S., Ishizaka, A., Labib, A., Saad, I.: Dominance-based rough set approach for group decisions. Eur. J. Oper. Res. **251**(1), 206–224 (2016)
6. Damart, S., Dias, L., Mousseau, V.: Supporting groups in sorting decisions: methodology and use of a multicriteria aggregation/disaggregation DSS. Decis. Support Syst. **43**(4), 1464–1475 (2007)
7. de Lima Silva, D., Ferreira, L., de Almeida-Filho, A.: A new preference disaggregation TOPSIS approach applied to sort corporate bonds based on financial statements and expert's assessment. Expert Syst. Appl. **152**(2), 113369 (2020)
8. Dias, L., Mousseau, V.: IRIS: a DSS for multiple criteria sorting problems. J. Multi-Criteria Decis. Anal. **12**, 285–298 (2003)
9. Dias, L., Mousseau, V.: Inferring ELECTRE's veto-related parameters from outranking examples. Eur. J. Oper. Res. **170**(1), 172–191 (2006)
10. Dias, L., Mousseau, V., Figueira, J., Clímaco, J.: An aggregation/disaggregation approach to obtain robust conclusions with ELECTRE TRI. Eur. J. Oper. Res. **138**, 332–348 (2002)
11. Doumpos, M., Zopounidis, C.: Preference disaggregation for multicriteria decision aiding: an overview and perspectives. In: Doumpos, M., Figueira, J.R., Greco, S., Zopounidis, C. (eds.) New Perspectives in Multiple Criteria Decision Making. MCDM, pp. 115–130. Springer, Cham (2019). https://doi.org/10.1007/978-3-030-11482-4_4
12. Doumpos, M., Zopounidis, C.: Disaggregation approaches for multicriteria classification: an overview. In: Matsatsinis, N., Grigoroudis, E. (eds.) Preference Disaggregation in Multiple Criteria Decision Analysis. MCDM, pp. 77–94. Springer, Cham (2018). https://doi.org/10.1007/978-3-319-90599-0_4

13. Figueira, J.R., Mousseau, V., Roy, B.: ELECTRE methods. In: Greco, S., Ehrgott, M., Figueira, J.R. (eds.) Multiple Criteria Decision Analysis. ISORMS, vol. 233, pp. 155–185. Springer, New York (2016). https://doi.org/10.1007/978-1-4939-3094-4_5
14. Greco, S., Ehrgott, M., Figueira, J. (eds.): ELECTRE Methods. Springer, New York (2016)
15. Greco, S., Kadzinski, M., Mousseau, V., Slowinski, R.: Robust ordinal regression for multiple criteria group decision: UTA^{GMS}-GROUP and UTADIS^{GMS}-GROUP. Decis. Support Syst. **52**(3), 549–561 (2012)
16. Kadziński, M., Martyn, K., Cinelli, M., Słowiński, R., Corrente, S., Greco, S.: Preference disaggregation method for value-based multi-decision sorting problems with a real-world application in nanotechnology. Knowl.-Based Syst. **218**, 106879 (2021)
17. Kolfschoten, G., French, S., Brazier, F.: A discussion of the cognitive load in collaborative problem-solving. EURO J. Decis. Process. **2**, 257–280 (2014)
18. Madhooshiarzanagh, P., Abi-Zeid, I.: A disaggregation approach for indirect preference elicitation in Electre TRI-nC: application and validation. J. Multi-Criteria Decis. Anal. (2021, in press). https://doi.org/10.1002/mcda.1730
19. Mousseau, V., Slowinski, R.: Inferring an ELECTRE TRI model from assignment examples. J. Global Optim. **12**, 157–174 (1998)
20. Ngo The, A., Mousseau, V.: Using assignment examples to infer category limits for the ELECTRE TRI method. J. Multi-Criteria Decis. Anal. **11**(2), 29–43 (2002)
21. Perry, J., Powers, R.: Aggregation rules that satisfy anonymity and neutrality. Econ. Lett. **100**(1), 108–110 (2008)
22. Waegeman, W., De Baets, B., Boullart, L.: Kernel-based learning methods for preference aggregation. 4OR **7**(2), 169–189 (2009)

Machine Learning, Recommender Systems, and Knowledge Systems

A Dynamic Convolutional Neural Network Approach for Legal Text Classification

Eya Hammami[1]([✉]), Rim Faiz[2], and Imen Akermi[1]

[1] LARODEC Laboratory, University of Tunis, Tunis, Tunisia
imen.akermi@isg.rnu.tn
[2] LARODEC Laboratory, University of Carthage, Carthage, Tunisia
rim.faiz@ihec.rnu.tn

Abstract. The Amount of legal information that is being produced on a daily basis in courts is increasing enormously. The processing of such data has been receiving considerate attention thanks to their availability in an electronic form and the progress made in Artificial Intelligence application. Indeed, deep learning has shown promising results when used in the field of natural language processing (NLP). Neural Networks such as convolutional neural networks and recurrent neural network have been used for different NLP tasks like information retrieval, sentiment analysis and document classification. In this work, we propose a Neural Network based model with a dynamic input length for French legal text classification. The proposed approach, tested over real legal cases, outperforms baseline methods.

Keywords: Natural language processing · Document categorization · Legal domain · Artificial intelligence

1 Introduction

The continued application of computational intelligence in legal domain has drawn a lot of attention in the last few decades. With the increased availability of legal text in digital form a wide variety of applications, including summarization [1], reasoning, classification [2], translation, text analytics, and others have been addressed within the legal domain. In this paper we particularly tackle text classification. Indeed, there are several applications that require partitioning natural language data into groups, e.g. classifying opinions retrieved from social media sites, or filtering spam emails, etc. In this work, we assert that law professionals would considerably gain advantage from the potential supplied by machine learning. This is especially the case for law professionals who have to take complicated decisions regarding several aspects of a given case. Given data accessibility and machine learning techniques, it is possible to train text categorization systems to predict some of these decisions. Such systems can act as a

ICIKS 2021.

© Springer Nature Switzerland AG 2021
I. Saad et al. (Eds.): ICIKS 2021, LNBIP 425, pp. 71–84, 2021.
https://doi.org/10.1007/978-3-030-85977-0_6

decision support system for law professionals. Several approaches have been proposed for text classification to mention, Naive Bayes classifier, Support Vector Machine, Logistic Regression, and most recently deep learning methods such as Convolutional Neural Network (CNN) [3,4], Recurrent Neural Network (RNN) and Long-Short Term Memory (LSTM) [5]. Most of these approaches are not particularly designed for the legal domain and are usually trained with English text, which make them not appropriate to be used for French text and particularly legal French text. Indeed, French is a language with a richer morphology and a more flexible word order, that requires more preprocessing to achieve good accuracy results and capture the hidden semantics specially when dealing with legal texts. In this paper, we propose NN-based model with dynamic input length layer to process French legal data. We also present a comparative study between the proposed approach and several baseline models. This paper is organized as follows: we present in Sect. 2 a literature review that examines the different approaches for text classification. In Sect. 3, we describe our proposed model. Finally, experiments and the deployment part are presented in Sect. 4.

2   Related Work

This section presents a brief discussion on the text classification task and on the application of deep learning to legal domain which includes various models developed for retrieving and classifying relevant legal text. Text classification is a necessary task in Natural Language Processing. Linear classifiers were frequently used for text classification [6,7]. According to [8] these linear models could scale up to a very huge dataset rapidly with a proper rank constraint and a fast loss approximation. Deep learning methods, such as recurrent neural networks (RNN) and Long Short Term Memory (LSTM) have been widely used in language modeling. Those methods are adapted to natural language processing because of their ability to extract features from sequential data, to mention the Convolutional Neural Network (CNN) which is usually used for computer vision tasks like in [9–11]. This model has been adopted in NLP for the first time in [12]. In this work the authors presented a new global max-pooling operation, which was revealed to be efficient for text, as an alternative to the conventional local max-pooling of the original LeNet architecture [13]. Furthermore, they suggested to transfer task-specific information by co-training different deep models on many tasks.

Inspired by the original work of [9,12] introduced a simpler architecture with modifications consisting of fixed pre-training word2vec embeddings. They proceed that both multitask learning and semi-supervised learning enhance the generalization of the shared tasks, resulting in state-of the-art-performance. Moreover, in [14], the authors demonstrated that this model can actually achieve state-of-the-art performances on many small datasets. Dynamic Convolutional Neural Network (DCNN) is a type of CNN which is introduced by [15]. Their approach outperforms other methods on sentiment classification. They use a new pooling layer called a dynamic K-max pooling. This dynamic k-max pooling is a generalization of the max pooling operator, which computes a new adapted K value for each iteration. Thus, their network can read any length of an input. Character-level Convolutional Neural Network (Char-CNN) which is introduced by [10] also yields better results than other methods on sentiment analysis and text classification. In the same context, [16] shows that a character-based embedding in CNN is an effective and efficient technique for sentiment analysis that uses less learnable parameters in feature representation. Their proposed method performs sentiment normalization and classification for unstructured sentences. A new Char-CNN model proposed by [17] and inspired from the work presented in [10], allows any length of input by employing k-max pooling before a fully connected layer to categorize Thai news from a newspaper. Furthermore, the work in [18] presented a character aware neural language model by combining a CNN on character embeddings with a Highway-LSTM on subsequent layers. Their results suggest that on many languages, character inputs are relevant for language modeling. In addition, [19] analyzed a multiplicative LSTM (mLSTM) on character embeddings and found out that a basic logistic regression learned on this representation can reach state-of-the art results on the Sentiment TreeBank dataset [20] with a few hundred labeled examples. We have noticed a rather small body of previous works about automatic text classification for legal documents. For example, support vector machines (SVMs) have been used to classify legal documents like legal docket entries in [21]. The authors developed simple heuristics to address the conjunctive and disjunctive errors of classifiers and improve the performance of the SVMs. Based on the prescience gained from their experiments, they also developed a simple propositional logic based classifier using hand labeled features, that addresses both types of errors simultaneously. They proved that this simple, approach outperforms all existing state-of-the-art ML models, with statistically significant gains. A mean probability ensemble system combining the output of multiple SVM classifiers to classify French legal texts, was also developed by [22]. They reported accuracy scores of 98% for predicting a case ruling, 96% for predicting the law area of a case, and 87.07% on estimating the date of a ruling. A preliminary study addressing deep learning for text classification in legal documents was proposed in [23]. They compered deep learning results with results obtained using SVM algorithm on four datasets of real legal documents. They demonstrated that CNN present better accuracy score with a training dataset of larger size and can be improved for text classification in the legal domain. Neural Networks such as CNN, LSTM and RNN have also been used for classifying English legal court opinions of Washington University School of Law Supreme Court Database (SCDB) in [24]. The authors compared the machine learning algorithms with several Neural Networks systems and they found out that CNN combined with Word2vec performed better compared to the other approaches and gave an accuracy around 72.7%. Based on the Brazilian Court System representing the biggest judiciary system in the world, and receiving an extremely high number of lawsuits every day, the work in [25] presented a CNN based approach that helps analyse and classify these cases, in order to be associated to relevant tags and allocated to the right team. The obtained results are very promising. However, most of the mentioned approaches are generally based on the CNN model and usually use a static input length. Therefore, we propose to experiment this model with a dynamic input length on French legal data. Experiments on real datasets highlight the relevance of our approach and open up many perspectives.

3 Proposed Model

The architecture of our proposed model, shown in Fig. 1, is based on the CNN model [24], characterized by a max pooling layer also called temporal max pooling, which is a method for down sampling data by utilizing a gliding window on a row of data and choosing a cell which includes a maximum value to be moved to the next layer. It carries out an operation on 1D CNN and it is calculated by the following formula (1) [17]:

$$P_{r,c} = max_{j=1}^{s} M_{r,s(c-1)+j} \qquad (1)$$

where:

– M is an input matrix with a dimension of $n \times l$
– s is a pooling size
– P is an output matrix with a dimension of $n \times \frac{l}{s}$
– c is a column cell of matrix P
– r is a row cell of matrix P

It is within this pooling layer that we try to experiment pooling layer, thus using the k-max pooling layer rather than the max-pooling layer. In fact the k-max pooling operation enables to pool the k maximum active features in P, also, it keeps the order of the features, but it is insensitive to their accurate positions. It can then detect more delicately the number of times where the feature is activated in P. The k-max pooling operator is used in the network after the highest convolutional layer. This allows the input to the fully connected layers to be separate from the length of the input sentence. Additionally, in the middle of the convolutional layers, the pooling parameter k is not fixed, but is selected in a dynamic way to enable a sleek extraction of a longer-range and higher order features [15]. This dynamic pooling layer is calculated by the following formula (2) [17]:

$$P_{r,*} = kmax_{j=1}^{l} M_{r,j} \qquad (2)$$

where:

– M is an input matrix with a dimension of $n \times l$
– K is an integer value
– P is an output matrix with a dimension of $n \times k$
– $*$ shows that all columns in a row are calculated together
– r is a row cell of matrix P

The main difference between these two types of pooling layer consists in the use of a gliding window. Max pooling is a method for down sampling data by utilizing a gliding window on a row of data and choosing a cell which includes a maximum value to be moved to the next layer [17]. Differently, k-max pooling doesn't have a window, but it has a choosing operation which carries out all data in a row. Then, top k cells which have maximum value are chosen to be utilized in the up-coming layer [17]. By applying K-max pooling in a convolutional neural network, as we propose, we can definitely have a matrix which is able to fit into a fully connected layer regardless of the length of an input. Figure 1 illustrates in details our proposed method.

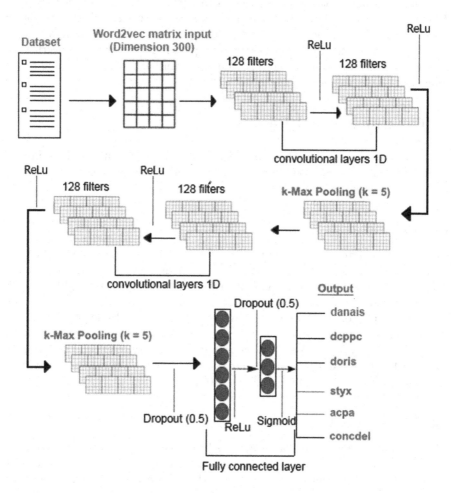

Fig. 1. Proposed architecture.

On the convolutional and pooling layers, the data length belongs to the input. Whereas, after the k-max pooling layer, the data length in each document is coequal. Thus, our neural network classification model is a little bit similar to the one introduced by [9], but we modified the layers, by adding other layers and modifying some of the original hyperparameters in order to obtain a better performed text categorization model. Our model first makes an embedding layer using word2vec as a pre-trained word embedding, and next makes a matrix of documents represented by 300-dimensional word embedding. As we all know when employing machine learning methods in NLP, most of the studies use 200 or 300 dimensional vectors, but 300-dimensional embedding carry more information and this, therefore, is considered to produce better performance results. Then, we incorporate the following sets of parameters: A dropout of 0.5, because it helps to change the concept of learning all the weights together to learning a fraction of the weights in the network in each training iteration; a convolution layer of 128 filters with a filter size of 3, according to the literature, we set the k value to 5. We also add a dense layer consisting of 128 units between two dropouts of 0.5 to prevent overfitting. Finally, the last layer (output layer) is a dense layer with a size of 6 equal to the number of labels (categories) in our dataset.

4   Experiments and Results

We present in this section the experimental results of our approach compered to the different methods used in the literature. We use the accuracy and the F_1 scores to evaluate these models.

4.1   Dataset

We trained and tested our model on a French legal dataset collected from data.gouv.fr[1]. It is a documentary collection of lawsuits from French courts. The dataset includes 2000 documents (txt files). The following figure (Fig. 2) presents a sample of this dataset. This dataset is organized into 6 categories whose denomination were carried out by a legal expert (see Table 1). The number of documents was limited because the documents annotation is done manually and exclusively by legal experts. Work is underway to try to expand the training corpus. After processing, the vocabulary size is 3794659. We randomly divide it into training and test set, with 80% and 20% split.

4.2   Pre-processing

Our model first removes special characters like punctuation, stopwords, numbers and whitespaces. The removal of these special characters will allow us to have classes that are representative of the words that are recurring in our documents. Second, we proceed with lemmatization by using TreetaggerWrapper module and removing named entities after recognize them using French Spacy and NLTK

---

[1] www.data.gouv.fr/fr/.

CAAIX1416771 - Bloc-notes

Fichier   Edition   Format   Affichage   Aide

Cour d'appel, Aix-en-Provence, 10e chambre, 4 Mai 2016 - n° 14/16771
Cour d'appel
Aix-en-Provence
10e chambre
4 Mai 2016
Numéro de rôle: 14/16771
Numéro: 2016/204
X / Y
Contentieux Judiciaire
COUR D'APPEL D'AIX-EN-PROVENCE
10e Chambre
ARRÊT AU FOND
DU 04 MAI 2016
N° 2016/204
Rôle N° 14/16771
Pasquale D.
C/
Thierry H.
CAISSE PRIMAIRE D'ASSURANCE MALADIE DES ALPES MARITIMES
SA FCA MOTOR VILLAGE FRANCE
Grosse délivrée
le:
à:
Me Franck G.
SCP M. -V.-M.,
Me Vincent P.
Décision déférée à la Cour:
Jugement du Tribunal d'Instance de NICE en date du 17 Juillet 2014 enregistré au
répertoire général sous le n° 12-000238.
APPELANT
Monsieur Pasquale D.
né le 28 Mars 1970 à [...], de nationalité Française,
demeurant [...]
représenté par Me Franck G., avocat au barreau de GRASSE
INTIMES
Monsieur Thierry H.

Fig. 2. Document sample.

Table 1. Predefined categories.

Number of documents	Label
298	DANAIS
501	DCPPC
159	DORIS
160	STYX
582	CONCDEL
300	ACPA

modules to allow a more accurate interpretation of the data. We chose to per-form lemmatization rather than stemming because lemmatization considers the context and converts the word to its meaningful base form, whereas stemming just removes the last few characters, often leading to incorrect meanings and spelling errors. Finally, each word in the corpus is mapped to a word2vec vector before being fed into the convolutional neural network for categorization.

4.3   Experiments

In this paper, we use Accuracy as a measure of evaluation in order to deter-mine the degree of predictions that the models was able to guess correctly. It is calculated like the following (3):

$$Accuracy = \frac{number\,of\,correctly\,classified\,documents}{total\,number\,of\,classified\,documents} \tag{3}$$

We also consider the F_1 score (4) which is a metric that combines both Precision and Recall using the Harmonic mean. In this work, our classification problem based on imbalanced class distribution, thus F_1 score is a better metric to evaluate our model on. $F_{1,i}$ refers to F_1 of class i, C is the number of categories:

$$F_1 = \frac{\sum_{i=1}^{C} F_{1,i}}{|C|} \tag{4}$$

Where:

$$F_{1,i} = 2 \cdot \frac{precision_i \cdot recall_i}{precision_i + recall_i} \tag{5}$$

Along with our CNN K-max pooling approach, we experimented two other CNN based models: CNN max-pooling and CNN global max-pooling. In the CNN with max pooling, we use the same hyperparameters as the CNN with global max pooling, but we change the pooling size to 3. The implementation of these three architectures is done using Keras which allows users to choose whether the models they build are running on Theano or TensorFlow. In our case the models run on TensorFlow.

Regularization of Hyperparameters: In our experiments, we tested our model with a set of various hyperparameters. The model performed best when using 128 filters for each of the convolutional layers. In addition, each of the models is adjusted with a dropout [27], which works by "dropping out" a pro-portion p of hidden units throughout training. We found out that a dropout of 0.5 and a batch size of 256 worked best for our CNNs models, along with the Adam optimizer [26].

Results of the First Experiment (K = 5): As shown in Table 2, CNN with max pooling performs better. It can achieve an accuracy of 84,46 %, which outperforms the CNN with k-max pooling (our proposed approach) and the CNN with global max pooling. We think that this is maybe due to the limits number of documents in the dataset.

Table 2. Results of the first experiments where k value set to 5.

Method	Accuracy (%)	F1 (%)
CNN with k-max pooling	80.35	80.20
CNN with max pooling	**84,46**	**84.46**
CNN with global max pooling	81,94	82.10

Results of the Second Experiment (K = 3): Now we decrease the k value to 4 then to 3. The purpose of this second experiment was explore if we could get a better accuracy when varying K with the proposed K-max pooling approach. As shown in Table 3, our model outperforms the other models with 84.71% accuracy when K is set to 3.

Table 3. Results of the second experiments where k value set to 3 which outperforms the other models.

Method	Accuracy (%)	F1 (%)
CNN with k-max pooling (K = 3)	**84.71**	**84.80**
CNN with k-max pooling (K = 4)	83,32	83.11
CNN with max pooling	84,46	84.46
CNN with global max pooling	81,94	82.10

As follows we present the two plots of accuracy and Cross Entropy for this second experiment: Fig. 3 shows the regression of the Cross Entropy and the evolution of the accuracy for CNN with k-max pooling (where k = 3) according to the number of epochs for both training and test sets. The red curve corresponds to the validation and the blue curve corresponds to the training. In this two graphics we notice that the line plot is well converged for the two curves and gives no sign of over or under fitting.

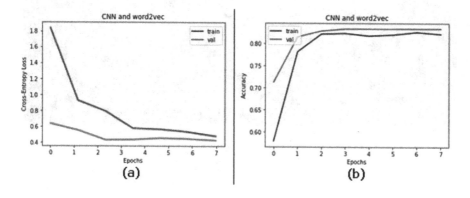

Fig. 3. Line plot of Cross Entropy Loss over Training Epochs.

Comparison with Baseline Methods: We also compared the CNN based models to three traditional methods: Naive Bayes Classifier (with TF-IDF) [28], Word2vec embedding with Logistic Regression [29] and SVM [21]. The results are shown in Table 4 as we can notice, CNN with k-max pooling outperforms non NN based models.

Table 4. Results of comparison with baseline methods.

Method	Accuracy (%)	F1 (%)
CNN with k-max pooling	**84.71**	**84.80**
CNN with max pooling	84,46	84.46
CNN with global max pooling	81,94	82.10
Naive Bayes classifier	41,91	42,00
Word2vec and Logistic Regression	80,88	80,01
SVM	79,98	79,51

Deployment: We developed a small desktop application based on our proposed method, which is designed for law professionals to allow them to categorize automatically textual data.

Figure 4 presents a screenshot of the home interface and Fig. 5 shows how they can easily load a simple French legal txt file to predict its classification according to predefined categories (see Fig. 6).

However, We are currently working on enhancing this application in order to integrate more functionalities that can help law professionals with heavy manual tasks.

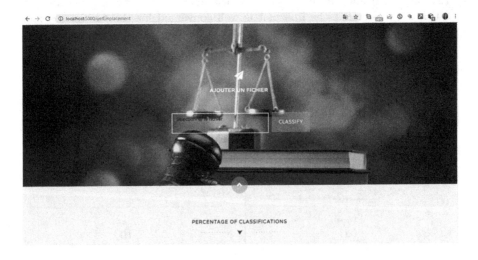

Fig. 4. The application Home page.

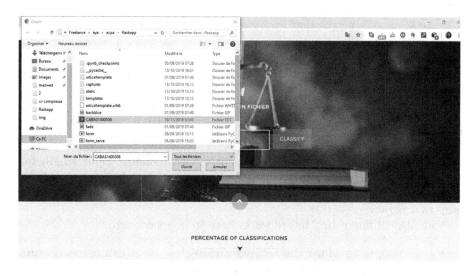

Fig. 5. Load a txt file.

Fig. 6. Percentages of a txt file classification.

5   Discussion

Dynamic max pooling [17], usually proved to perform much better compared to classic max pooling and other baseline methods. But in our first experiment, the static max pooling outperforms all other methods with an accuracy of 84,46%. Then in our second experiment, we adjusted the K value of the Dynamic max pooling to 3, as a result the Dynamic max pooling gets better accuracy result of 84,71%. By the way, it outperforms all other methods. In this work we considered that with dynamic k-max pooling, the value of k depends on the input shape. The idea is that longer sentences can have more max values (higher k). But in our case, the sentence's length that we have, are not enough to set higher K value. We considered also that the words contained in the pre-trained word embedding may not capture the specificity of languages in legal domain. Therefore, we think maybe for these reasons our results may not be very optimal especially for the first experiment.

6 Conclusion

In this paper, we addressed the use of CNN with dynamic input length for French legal data classification. Our suggested approach, which can process a longer input length, outperforms the original model with a fixed input length in terms of accuracy.

A number of interesting future works have to be mentioned:

Firstly, we plan to re-adjust the network architecture, so it can better capture the characteristics of our French legal data.

Secondly, we should test our proposed approach on new datasets to validate its performance.

Finally, We think that we can extend our reflections to the categorization of hand written documents and not be limited to electronic versions.

Acknowledgments. This paper has been done under the contract PREMATTAJ 2017–2019 of the Occitanie region which is greatly acknowledged. The decisions used in this paper have been annotated by Professor Guillaume Zambrano of the University of Nîmes.

References

1. Doan, T.M., Jacquenet, F., Largeron, C., Bernard, M.: A study of text summarization techniques for generating meeting minutes. In: Dalpiaz, F., Zdravkovic, J., Loucopoulos, P. (eds.) RCIS 2020. LNBIP, vol. 385, pp. 522–528. Springer, Cham (2020). https://doi.org/10.1007/978-3-030-50316-1_33
2. Stead, C., Smith, S., Busch, P., Vatanasakdakul, S.: Towards an academic abstract sentence classification system. In: Dalpiaz, F., Zdravkovic, J., Loucopoulos, P. (eds.) RCIS 2020. LNBIP, vol. 385, pp. 562–568. Springer, Cham (2020). https://doi.org/10.1007/978-3-030-50316-1_39
3. Li, Y., Hao, Z.B., Lei, H.: Survey of convolutional neural network. J. Comput. Appl. **36**, 2508–2515 (2016)
4. Albawi, S., Mohammed, T.A., Al-Zawi, S.: Understanding of a convolutional neural network. In: International Conference on Engineering and Technology, ICET, pp. 1–6 (2017)
5. Hochreiter, S., Schmidhuber, J.: Long short-term memory. Neural Comput. **9**, 1735–1780 (1997)
6. Joachims, T.: Text categorization with Support Vector Machines: learning with many relevant features. In: Nédellec, C., Rouveirol, C. (eds.) ECML 1998. LNCS, vol. 1398, pp. 137–142. Springer, Heidelberg (1998). https://doi.org/10.1007/BFb0026683

7. McCallum, A., Nigam, K.: A comparison of event models for Naive Bayes text classification. In: AAAI-98 Workshop on Learning for Text Categorization, pp. 41–48 (1998)

8. Joulin, A., Grave, E., Bojanowski, P., Mikolov, T.: Bag of tricks for efficient text classification. In: Proceedings of the 15th Conference of the European Chapter of the Association for Computational Linguistics, pp. 427–431 (2017)

9. Kim, Y.: Convolutional neural networks for sentence classification. In: Proceedings of the 2014 Conference on Empirical Methods in Natural Language Processing (EMNLP), pp. 1746–1751 (2014)

10. Zhang, X., Zhao, J., LeCun, Y.: Character-level convolutional networks for text classification. In: Proceedings of the 29th Conference on Neural Information Processing Systems, NIPS 2015. Advances in Neural Information Processing Systems, pp. 649–657 (2015)

11. Conneau, A., Schwenk, H., Barrault, L., Lecun, Y.: Very deep convolutional networks for text classification. In: Proceedings of the 15th Conference of the European Chapter of the Association for Computational Linguistics, pp. 1107–1116 (2017)

12. Collobert, R., Weston, J.: A unified architecture for natural language processing: deep neural networks with multitask learning. In: Proceedings of the 25th International Conference on Machine Learning, ICML, pp. 160–167 (2008)

13. LeCun, Y., Bottou, L., Bengio, Y., Haffner, P.: Gradient-based learning applied to document recognition. In: Proceedings of the IEEE, pp. 2278–2324 (1998)

14. Mikolov, T., Chen, K., Corrado, G., Dean, J.: Efficient estimation of word representations in vector space. In: International Conference on Learning Representations (2013)

15. Kalchbrenner, N., Grefenstette, E., Blunsom, P.: A convolutional neural network for modelling sentences. In: Proceedings of the 52nd Annual Meeting of the Association for Computational Linguistics, pp. 655–665 (2014)

16. Arora, M., Kansal, V.: Character level embedding with deep convolutional neural network for text normalization of unstructured data for Twitter sentiment analysis. Soc. Netw. Anal. Min. **9**(1), 1–14 (2019). https://doi.org/10.1007/s13278-019-0557-y

17. Koomsubha, T., Vateekul, P.: A character-level convolutional neural network with dynamic input length for Thai text categorization. In: Proceedings of the 9th International Conference on Knowledge and Smart Technology, KST, pp. 101–105 (2017)

18. Kim, Y., Jernite, Y., Sontag, D., Rush, A.M.: Character-aware neural language models. In: Proceedings of the Thirtieth AAAI Conference on Artificial Intelligence (2016)

19. Radford, A., Jozefowicz, R., Sutskever, I.: Learning to generate reviews and discovering sentiment. In: International Conference on Learning Representations, ICLR (2018)

20. Socher, R., et al.: Recursive deep models for semantic compositionality over a sentiment treebank. In: Proceedings of Conference on Empirical Methods in Natural Language Processing, pp. 1631–1642 (2013)

21. Nallapati, R., Manning, C.D.: Legal docket classification: where machine learning stumbles. In: Proceedings of the 2008 Conference on Empirical Methods in Natural Language Processing, pp. 438–446 (2008)

22. Sulea, O.M., Zampieri, M., Malmasi, S., Vela, M., Dinu, L.P., van Genabith, J.: Exploring the use of text classification in the legal domain. In: Proceedings of 2nd Workshop on Automated Semantic Analysis of Information in Legal Texts, ASAIL (2017)

23. Wei, F., Qin, H., Ye, S., Zhao, H.: Empirical study of deep learning for text classification in legal document review. In: International Conference on Big Data, pp. 3317–3320 (2018)
24. Undavia, S., Meyers, A., Ortega, J.E.: A comparative study of classifying legal documents with neural networks. In: Federated Conference on Computer Science and Information Systems, FedCSIS, pp. 515–522 (2018)
25. Da Silva, N.C., et al.: Document type classification for Brazil's supreme court using a convolutional neural network. In: Proceedings of the Tenth International Conference on Forensic Computer Science and Cyber Law-ICoFCS, pp. 7–11 (2018)
26. Kingma, D.P., Ba, J.: Adam: a method for stochastic optimization. In: International Conference on Learning Representations (2015)
27. Srivastava, N., Hinton, G., Krizhevsky, A., Sutskever, I., Salakhutdinov, R.: Dropout: a simple way to prevent neural networks from over fitting. J. Mach. Learn. Res. JMLR 1929–1958 (2014)
28. Yoo, J.-Y., Yang, D.: Classification scheme of unstructured text document using TF-IDF and Naive Bayes classifier. The Journal of Machine Learning Research, Proceedings of 3rd International Conference on Computer and Computing Science, COMCOMS (2015)
29. Pranckevičius, T., Marcinkevičius, V.: Comparison of Naive Bayes, random forest, decision tree, support vector machines, and logistic regression classifiers for text reviews classification. Baltic J. Mod. Comput. 5, 221 (2017)

Trends of Evolutionary Machine Learning to Address Big Data Mining

Sana Ben Hamida[1](✉), Ghita Benjelloun[2], and Hmida Hmida[3]

[1] Université Paris Dauphine, PSL Research University, CNRS,
UMR[7243], LAMSADE, Paris, France
`Sana.mrabet@dauphine.psl.eu`
[2] Université Paris Dauphine, PSL Research University, Paris, France
[3] Université de Tunis El Manar, Tunis, Tunisia

Abstract. Improving decisions by better mining the available data in an Information System is a common goal in many decision making environments. However, the complexity and the large size of the collected data in modern systems make this goal a challenge for mining methods. Evolutionary Data Mining Algorithms (EDMA), such as Genetic Programming (GP), are powerful meta-heuristics with an empirically proven efficiency on complex machine learning problems. They are expected to be applied to real-world big data tasks and applications in our daily life. Thus, they need, as all machine learning techniques, to be scaled to Big Data bases. This paper review some solutions that could be applied to help EDMA to deal with Big Data challenges. Two solutions are then selected and explained. The first one is based on the algorithmic manipulation involving the introduction of the active learning paradigm thanks to the active data sampling. The second is based on the processing manipulation involving horizontal scaling thanks to the processing distribution over networked nodes. This work explains how each solution is introduced to GP. As preliminary experiences, the extended GP is applied to solve two complex machine learning problem: the Higgs Boson classification problem and the Pulsar detection problem. Experimental results are then discussed and compared to value the efficiency of each solution.

Keywords: Big data mining · Machine learning · Genetic Programming · Horizontal parallelization · Active learning · Data sampling · Higgs Boson classification · Pulsar detection

1 Introduction

Modern processes are frequently monitored using information systems that record large amounts of information given rise to Big Data bases. For a decision making environment, better mining the available data is a challenge. It is the major issue for the current data mining and machine learning algorithms. Regardless the Big Data 3V (variety, velocity and volume), data mining requires powerful knowledge discovery techniques that are able to deal with the related

© Springer Nature Switzerland AG 2021
I. Saad et al. (Eds.): ICIKS 2021, LNBIP 425, pp. 85–99, 2021.
https://doi.org/10.1007/978-3-030-85977-0_7

challenges such complexity and learning cost. Thus, the use of evolutionary data mining procedures becomes a hot trend. Thanks to their global search in the solution space, evolutionary computation techniques help in the information retrieval from a voluminous pool of data in a better way compared to traditional retrieval techniques [10].

Genetic Programming (GP) [21], as a particular EDMA, is considered as a universal machine learning tool. Its paradigm has shown a great potential when applied to supervised learning, especially classification and regression problems [3,32]. However, like other machine learning techniques, GP computational overhead increases with large datasets and its efficiency is affected. Some improvements are needed to scale GP when learning from big datasets. The adaptation of EDMA, specially GP, for big data problems may require redesigning the algorithms and/or their inclusion in parallel environments. Both tracks are discussed in this paper.

An efficient strategy to redesign a machine learning technique to handle large data sets is to introduce a special learning paradigm or a special parallelization paradigm without parallel hardware. In the first case, the algorithm of the data mining method is modified in order to improve its efficiency. With the second case, a parallel data-intensive computing scenario is introduced to the method. This strategy has become very popular in the last few years thanks to the appearance of some open-source tools such as Hadoop/MapReduce.

As discussed in [25], learning paradigms could bring some solutions for big data mining. For example, Ensemble Learning can help alleviate the "Concept Drift" and "Curse of Modularity" issues associated with Big Data. Similarly, Local Learning can help alleviate "Data locality" and "Variance and Bias" issues. In this work, we are interested in the Active Learning paradigm. Active learning is implemented essentially with active data sampling. In previous works [15,17], we demonstrated how active sampling could be an efficient solution to learn from large data sets by decreasing the size of the training set. The idea is to introduce some components in the algorithm that allows it to select its training data according to the evolution of the learning process. Section 4 explains in details how an active sampling can be introduced in the GP engine.

The second strategy to scale an EA (Evolutionary Algorithm) to Big Data Mining is the parallelization in a distributed environment. Vertical parallelization (scaling up) paradigm was widely explored. It includes multicore CPUs, supercomputers, hardware acceleration including graphic processing units (GPUs) and field-programmable gate arrays (FPGAs) [25]. In this work, we are interested in the horizontal parallelization (scaling out) paradigm where processing is dispersed over networked nodes. This paradigm do not require detailed knowledge of the underlying hardware architecture. The most known framework implementing the horizontal parallelization is the MapReduce paradigm and its distributed file system [11], originally introduced by Google. However, MapReduce encounters difficulties when dealing with iterative algorithms. Several new frameworks have been proposed to provide better support for iterative processing such as

HaLoop [6] or Apache Spark[1]. It is this last framework that we propose to scale up GP to Big Data Mining with processing manipulation. Our goal is the implementation of GP in a distributed ecosystem by adapting existing libraries. In a previous work, we proposed a solution to port a known GP implementation to the Spark context in order to take the most of its proven potential [16]. The source code of this model is available at GitHub[2]. This implementation is reused in the present work.

The main purpose of this paper is to summarize the different strategies to address Big Data challenges with machine learning (Sect. 2). It then presents the horizontal parallelization and the active learning paradigm, two suited approaches for EDMA (Sects. 3 and 4). The general purpose and some implementation details of the proposed techniques in each strategy are then given and explained in details (Sects. 3 and 4). The efficiency of GP obtained with the different extensions are studied on the HIGGS classification problem and the pulsar detection problem in Sect. 5.

2   Some Approaches to Address Big Data Learning Cost

In modern Information Systems, the amount of stored data has reached a huge volume and their complexity has widely increased. New challenges had then arisen due to the data characteristics.

Approaches to countering the Big Data challenges for machine learning are quite diverse. They are summarized in [25] and classified according to the data analytics pipeline (see Fig. 1). Another review, published in [3], identifies methods and techniques to accelerate evolutionary algorithms when applied to learning tasks. Scaling approaches focus essentially on manipulating data, processing and algorithms. We summarize below the main categories of the different approaches and paradigms that can be developed to alleviate some issues associated with Big Data.

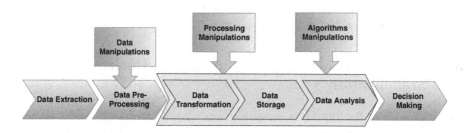

Fig. 1. Manipulations through Data Analytics pipeline [25].

---

[1] https://spark.apache.org.
[2] https://github.com/hhmida/gp-spark.

- Data manipulation: It is applied in the pre-processing phase before administering the data to the learning process. Its goal is to reduce the size of the learning data base, for example by applying techniques of sampling or feature selection. Sampling, in this case, is independent of the learning process (evolution process for GP).
- Processing manipulation: it relies on processing manipulations to handle the additional computational cost by vertical or horizontal scaling. It includes parallelization approaches using the parallel nature of some population based algorithms such as GP and EA. Parallelization can be done with:
  - Vertical scaling (scaling up): It is based on increasing resources for a single node. For example, hardware acceleration based on graphics processors called General Purpose Graphics Processing Units (GPGPU) and parallelization using multicore CPUs have been used in several jobs [14, 23, 27].
  - Horizontal scaling (scaling out): It refers to distributed systems where computations are deployed on a cluster of nodes [27]. This is the most common form of distribution with Big Data problems. Several frameworks have been put in place for two types of parallelization: by batch or flow-oriented. The most used are Hadoop/MapReduce [11] and Apache Spark [34].
- Algorithm Manipulation: it relies on algorithm manipulations in order to optimize its running time or to improve the learning quality. It involves the introduction of new machine learning paradigms or the adaptation of some existing machine learning paradigms. As new paradigms, the Online learning and Transfer learning are promising approaches. As existing paradigms that could be adapted to deal with Big Data, we find the Ensemble learning, the Local learning and the Active learning.

In the context of GP scaling, two paths are considered in this work: the acceleration of evaluations through the processing manipulation or the reduction of their number through the learning paradigm. In the first case, the GP engine is ported over a parallel framework to distribute the fitness computation. In the second case, the GP algorithm is extended with an active learning paradigm implemented with an active data sampling.

3   Scaling by Processing Manipulation: Horizontal Parallelization

The evolution of the Big Data ecosystem has allowed the development of new approaches and tools such as Hadoop MapReduce[3] and Apache Spark (see Footnote 1) that implement a new programming model over a distributed storage architecture. They are the de facto tools for any data intensive application. This section shows how an existing GP implementation is adapted to the Spark context.

---

[3] https://hadoop.apache.org.

3.1 Spark and MapReduce

MapReduce is a parallel programming model introduced by Dean et al. in [11] for Google. It was made popular with its Apache Hadoop implementation. The fundamental idea that gave rise to this model is to move computations to the data by reducing data traffic between the different nodes. MapReduce operates over a distributed file system.

It applies the "divide and rule" technique to break down a process into multiple tasks performed concurrently on different machines on the portions of data they store. These tasks are of two types: Map and Reduce, that are the functional programming origin. Despite the simplicity of this model, which has allowed to solve several large scale problems, it is not suited to iterative algorithms such as EDMA that are penalized by the large number of I/Os and network latency.

Apache Spark is one of many frameworks intended to neutralize the limitations of MapReduce while keeping its scalability, data locality and fault tolerance. The keystone of Spark is the Resilient Distributed Datasets (RDD) [34]. A RDD is a typical immutable parallel data structure that can be cached. These RDDs support two types of operations: transformations (*map. filter, join, ...*) that produce a modified RDD and actions (*count, reduce, ...*) that generate non-RDD-type results (an integer, a table, etc.) and require all the RDD partitions to be performed. Spark is up to 100 times faster than MapReduce with iterative algorithms (see Footnote 1).

3.2 Parallelizing GP over a Distributed Environment

The implementation of evolutionary algorithms over a distributed environment has taken up light since the emergence of the Big Data ecosystem. Several works have been published in this context. For example, Rong-Zhi Qi et al. [31] and Padaruru et al. [29] propose solutions for parallelizing the entire evolutionary process (fitness evaluation and genetic operators) with Spark. This solution has been applied to automatic software test game generation.

In Chávez et al. [7], the well-known EA library ECJ is modified in order to use MapReduce for fitness evaluations. This new tool is tested to resolve a face recognition problem over around 50 MB of data. Only time measure was considered in this work. Peralta et al. [30] applied the MapReduce model to implement an EA for the pre-processing of big datasets (up to 67 million records and attributes from 631 to 2000). It's a feature selection application where each map creates a vector of attributes on disjoint subsets of the original data base. The Reduce phase aggregates all the vectors obtained.

The implementation used in this paper, introduced in [16], is inspired by the works of Chavez et al. [7] Peralta et al. [30]. It is a solution for modifying an existing tool (DEAP) for the distribution of the training data base using the Spark engine for distributing GP evaluation. The GP loop with the different distribution steps are illustrated by the Algorithm 1. Steps 1, 4, 5, and 6 (blue lines) concern the distribution of the training data set for the population fitness computation. In step 1, we start by creating a RDD containing the training set. Then, at each generation, a map transformation (step 4) is performed by sending individuals code to be executed (step 5) on RDD and then get results to compute each individual fitness in step 6. Afterwards, GP pursues its standard evolutionary steps.

Algorithm 1. GP + RDD

 1: TrainingRDD ← parallelize the training set
 2: $g \leftarrow 0$
 3: **for all** generation $g < g_{max}$ **do**
 4:     map: serialize population and map it on the training set (trainingRDD)
 5:     TrainingRDD ← compute distributed fitness
 6:     Reduce : Compute final fitness (Reduce fitnessRDD)
 7:     Update population fitness
 8:     Select parents according to fitness
 9:     Apply genetic operators and generate new population
10: **end for**

---

4   Scaling by the Learning Paradigm: Active Learning

Active Learning [2,8] could be defined as: 'any form of learning in which the learning program has some control over the inputs on which it trains.'

Active learning is implemented essentially with active sampling techniques. The goal of any sampling approach is first to reduce the original size of the training set, and thus the computational cost, and second to enhance the learner performance. Sampling training dataset has been first used to boost the learning process and to avoid over-fitting [20]. Later, it was introduced for Genetic learners as a strategy for handling large input databases. With active sampling, the training subset is changed periodically across the learning process. The data selection strategy is based on some dynamic criteria, such as random selection, weighted selection, incremental selection, etc. We distinguish one-level sampling methods using a single selection strategy and multi-level sampling (hierarchical sampling) methods using multiple sampling strategies associated in a hierarchical way.

4.1   One-Level Active Sampling

One-level sampling methods use a single selection strategy based on dynamic criteria. Records in the training subset S are selected before the application of the genetic operators each generation.

To select a training subset S from a data base B, the simplest technique is to select randomly T_S records from B with a uniform probability. It is the basic approach for the Random Subset Selection method [13] (Sect. 4.1) and some variants like the Stochastic Sampling [28] and Fixed Random Selection [35]. Several other techniques use more sophisticated criteria in order to address some learning difficulties like over-fitting or imbalanced data. For example, weighted sampling techniques [13] use information about the current state of the training data such as difficulty and the number of solved fitness cases. However data-topology based sampling techniques [15, 24] use information about data topology in order to avoid to select similar fitness cases in the same training subset. Balanced sampling [19] and incremental sampling [22] techniques use information about the data classes to handle the problem of imbalanced data. For this work, we are interested in the random (RSS) and balanced (SBS) sampling methods. In this work, the training data base is divided on balanced blocks and SBS and RSS are applied the resulting blocks (Algorithm 2).

Random Sampling (RSS). The simplest method to choose fitness cases and build the training sample is random. The selection of fitness cases is based on a uniform probability among the training subset. This stochastic selection helps to reduce any bias within the full dataset on evolution. Random Subset Selection (RSS) is the first implementation given by Gathercole et al. [13]. In RSS, at each generation g, the probability of selecting any case i is equal to $P_i(g)$ such that:

$$\forall i : 1 \leq i \leq T_B, \quad P_i(g) = \frac{T_S}{T_B}. \tag{1}$$

where T_B is the size of the full dataset B and T_S is the target subset size. The sampled subset has a fluctuating size around T_S.

Balanced Sampling (SBS). The main purpose of balanced sampling is to overcome imbalance in the original data sets. The well known techniques in this category are those proposed by Hun et al. [19] aiming to improve classifiers accuracy by correcting the original dataset imbalance within majority and minority class instances. Some of these methods are based on the minority class size and thus reduce the number of instances. In this paper, we focus on the *Static Balanced Sampling (SBS)*. SBS is an active sampling method that selects cases with uniform probability from each class without replacement until obtaining a balanced subset. This subset contains an equal number of majority and minority class instances of the desired size.

4.2   Multi-level Active Sampling

Multi-level sampling combines several sampling algorithms applied at different levels. Its objective is to deal with large data sets that do not fit in the memory, and simultaneously provide the opportunity to find solutions with a greater generalization ability than those given by the one-level sampling techniques. The data subset selections at each level are independent. The usual schema is made up of three levels. The first one (level 0) consists in creating blocks with a given size from the original data set which are recorded in the hard disk. The remaining two levels are a combination of two active sampling methods. In the

---

Algorithm 2. GP + 2 Level Active Sampling

Parameters:

$\mathcal{B}$: learning data base

T_S: the subset size

 1: Divide $\mathcal{B}$ into blocks B_b *(level 0)*
 2: $\mathcal{S}(0) \leftarrow$ Select instances from B_b using RSS or SBS
 3: $g \leftarrow 0$
 4: **for all** generation $g < g_{max}$ **do**
 5:     Evaluate individuals against Training Subset $\mathcal{S}(g)$
 6:     Select parents according to fitness
 7:     Apply genetic operators and generate new population
 8:     Generate randomly new data subset $S(g+1)$ using RSS or SBS (level 1)
 9: **end for**

---

present work, a balanced sampling is applied at levels 0 and 1 and a random sampling is applied at level 2.

Algorithm 3 details how the hierarchical sampling is added to the GP loop. At level 0, the data set $\mathcal{B}$ is first partitioned into blocks that are selected randomly for the sampling step. Then, at level 1, an intermediate sample is extracted from the current block using a balanced random selection SBS. Finally, at level 2, the training subset $\mathcal{S}$ is generated randomly from the intermediate sample.

Algorithm 3. GP + 3 Level Active Sampling

Parameters:

$\mathcal{B}$: learning data base

T_{l1}: level 1 maximum iterations

T_{l2}: level 2 maximum iterations

 1: Divide $\mathcal{B}$ into blocks *(level 0)*
 2: **for all** level 1 iterations $= T_{l1}$ **do**
 3:     Conduct Block Selection
 4:     Sample an Intermediate subset using SBS or RSS *(level 1)*
 5:     **for all** level 2 iterations $= T_{l2}$ **do**
 6:         $\mathcal{S}(g) \leftarrow$ Conduct Training Subset Selection using RSS *(level 2)*
 7:         Evaluate individuals against Training Subset $\mathcal{S}(g)$
 8:         Select parents according to fitness
 9:         Apply genetic operators and generate new population
10:     **end for**
11: **end for**

In addition to the known GP parameters, two new parameters are used by the algorithm: T_{l1} and T_{l2}. T_{l1} and T_{l2} are respectively the number of GP iterations for the first-level sampling and for the second-level sampling. Thus, after each T_{l1} iterations, the level-one training data set is replaced with an other set with the SBS method. This set is used to generate the training sub set $\mathcal{S}$ at each T_{l2} iterations with the RSS approach.

5   Example of Application

This section presents an example of application of the proposed solutions in solving two real world problems. The aim is not to present a detailed experimental study but to give an example of the GP's behaviour when extended with an active sampling or a parallelization over Spark. For this purpose, we have chosen two applications: The Higg's Boson classification and the Pulsar detection problems.

5.1   Application Data Bases

The Higgs Data Base for Boson Detection. A Higgs or Z boson is a heavy state of matter resulting from a small fraction of the proton collisions at the Large Hadron Collider[5]. This heavy state quickly decays into more stable particles, so the intermediate states of matter are not observable by the detectors surrounding the point of collision. Highly faithful collisions are then simulated by the ATLAS Simulator [9] to reproduce the essential measurements provided by the detectors to reproduce the essential measurements provided by the detectors.

ATLAS experiment a portion of the simulated data to optimize the analysis of the Higgs Bosons by machine learning techniques. A subset of this data was presented as a challenge in 2014 [1] in the kaggle platform[4,5].

From the machine learning point of view, the problem can be formally cast into a binary classification problem. The task is to classify events as a signal (event of interest) or a background (event produced by already known processes). Baldi et al. [4] published for benchmarking machine-learning classification algorithms a big data set of simulated Higgs Bosons that contains 11 million simulated collision events and the 28 features. In this work, we propose to handle a subset of the published data to test th different GP implementations.

HTRU Data Set for Pulsar Detection. Pulsars are a rare type of Neutron star that produce radio emission detectable on Earth. They are of considerable scientific interest as probes of space-time, the inter-stellar medium, and states of matter [26]. As pulsars rotate, their emission beam sweeps across the sky, and when this crosses our line of sight, produces a detectable pattern of broadband radio emission. As pulsars rotate rapidly, this pattern repeats periodically. Thus pulsar search involves looking for periodic radio signals. The HTRU(2) dataset, is the publicly available data set[6] that describes a sample of pulsar candidates collected during the High Time Resolution Universe (HTRU) Survey. The goal of the classification process is to classify the given candidates as pulsars or non pulsars.

---

[4] https://www.kaggle.com/.

[5] See UCI Machine Learning Repository at http://archive.ics.uci.edu/ml/datasets/HIGGS.

[6] https://archive.ics.uci.edu/ml/datasets/HTRU2.

Some Related Works on Higgs and HTRU Data Sets. The first machine learning technique applied to detect pulsars on the Lyon et al. HTRU data set is the Gaussian Hellinger very fast Decision Tree with a precision equal to 89.9% [26]. These results are improved in the study published in [33] where the precision and recall reach respectively 95% and 87.3% with the XG-Boost classifier and 96% and 87% with Random Forest classifier.

Otherwise, the Boson classification problem using the Higgs data set was first handled with Deep Neural Network (DNN) proposed by Baldi et al.[5]. Their method was compared the boosted decision tree and the Shallow Neural Network. They used a subset of Higgs data base of 2.6 million examples and 100K validation examples. They demonstrated how DNN can be trained on such data set with a high degree of accuracy. The whole Kaggle data set with 11 millions patterns has been studied in [16,18] where the best accuracy reach 66.93%.

5.2   Experimental Settings

Software Framework. DEAP (Distributed Evolutionary Algorithms in Python)[7] is presented as a rapid prototyping and testing framework [12] that supports parallelization and multiprocessing. It implements several Evolutionary Algorithms: Genetic Algorithm, Evolution Strategies and GP. The basic module contains objects and data structures frequently used in Evolutionary Computation. We decided to use this framework for the following reasons: (1) it implements standard GP with tree based representation (1), it is a Python package which is one of the 3 languages supported by Spark and (3) it is distributed ready. The third point means that DEAP is structured in a way that facilitates distribution of computing tasks.

Configurations and Performance Metrics. For each data set, five series of tests are performed with different configurations according to the sampling strategy or parallelization strategy, as follows:

- Standard GP: GP is run without active learning or parallelization.
- GP + Spark: GP is parallelized over Spark.
- GP + 2 Level Sampling: GP is extended with a two level sampling using SBS at level 0 and RSS at level 1.
- GP + 3 Level Sampling: GP is extended with a three level sampling using SBS at level 0 and 1 and RSS at level 2.

By the end of each run, the best individual based on the fitness function is evaluated on the test data set. Results are recorded in a confusion matrix from which accuracy is calculated according to the following formula.

$$Accuracy = \frac{True\ Positives + True\ Negatives}{Total\ patterns}. \tag{2}$$

---

[7] https://github.com/deap/deap.

where *True Positives* and *True Negatives* are the numbers of exemplars correctly classified in respectively class 1 and 0. Experiments are performed on an Intel i7 (4 Core) workstation with 16 GB RAM running under *Windows* 64-bit Operating System.

The additional parameters needed for the different configurations are summarized in the Table 1. For these first experiments, the training size is limited 40000 for the case of Boson classification.

Table 1. Additional parameters for the GP runs.

Method	Parameter	Level 0	Level 1	Level 2
Standard GP and GP+Spark	Size	HIGGS: 40000 HTRU: 14000	– –	– –
GP+2 Level Sampling	Size Frequency	HIGGS: 40000 - HTRU:2000 50 g	5000 1	400 –
GP+3 Level Sampling	Size Frequency HTRU Frequency HIGGS	HIGGS:5000 - HTRU:2000 50 g 50 g	1000 5 8	200 2 4

GP Settings. The GP terminal set includes the features of data set benchmarks studied in this work. The GP function set includes basic arithmetic, comparison and logical operators reaching 17 functions. The objective is the minimisation of the classification error. The main GP parameters are summarized in Table 2.

Table 2. GP parameters.

Parameter	Value	Parameter	Value
Population size	200	Generations number	200
Crossover probability	0.5	Mutation probability	0.2
Tournament size	4	GP Tree depth	3

5.3   Results and Discussion

The results of these preliminary experiments are illustrated in Table 3 for the HTRU base and in Table 4 for the HIGGS data base. The recorded metrics are average and best accuracy and average computing time over 10 runs.

The key observation from the obtained results is that the proposed solutions are able not only to reduce the computation cost according to Standard GP results, but also to improve the performance of the classifiers. The reduction in the computational cost is about 50% with active learning and can be reduced

Table 3. Preliminary results on HTRU data set.

Test case	Avg accuracy	Computing time (en s)	Best accuracy
Standard GP	97%	5313,81	97,8%
GP+2L-Sampling	97,3%	764,183	**98,9%**
GP+3L-Sampling	97,2%	751,607	98,7%
GP+RDD	**97,4%**	**428,83**	98,2%

Table 4. Preliminary results on HIGGS data set.

Test case	Avg accuracy	Computing time (en s)	Best accuracy
Standard GP	57%	7918	59,2%
GP+2L-Sampling	53%	2220	56%
GP+3L-Sampling	52%	3417,89	55%
GP+RDD	**57%**	**1278,9**	**61%**

up to 10 times with GP+RDD in the case of HTRU. Likewise, the average and best accuracies are improved with parallelization and active learning in the majority of test cases. According to these first series of tests and considering the best measures, we can state that the extended GP is able to generate better results than the standard GP either for Higgs data set or HTRU data set.

Results for 1M Train Instances from Higgs Database. The GP behavior observed in the first experiments should remain unchanged with the increase of data size. To demonstrate it, some tests were carried out with a train sample of 1million instances from the Higgs data base and a test set of 200000 instances. Due to the limited computational capacity, only few runs are performed and the number of generations is limited to 25 for GP+Spark and to 100 for PG+2/3 L Sampling. The remaining GP parameters still unchanged.

GP+Spark, after 25 generations and about 5 h of computing time, the best test accuracy reach 57,15%. The parallelization of GP over Spark has not only allowed GP to be applied to a very large data set that is impossible to do with a Standard GP, but also to slightly improve the overall performance. For active learning, as for previous experiments, the training set is first divided into blocks of 100,000 instances (level 1). RSS is then applied on these blocks to generate samples with 10000 instances in the case of 2L sampling. For the 3L Sampling, SBS and RSS are applied respectively at level 1 and 2 where the sample sizes are equal of 10000 and 3000. The learning time reach 5h30mn, however the results are quite improved. The accuracy of the resulting classifiers is about 63% for 2L and 61% for 3L. This performance exceeds not only the first results in Table 4 but also some previous published results in the state of the art. However, it is not possible to compare the results of this paper with those of baldi et al. [5] since we use a smaller sample size.

Discussion. This paper presents some trends to address difficulties related to complexity and volume for big data mining with GP as an EDMA. Two solutions are explored, the active sampling with SBS and RSS and the horizontal parallelization with Spark, that are proved to be promising directions to improve classifiers' quality. Indeed, according to the first series of tests summarized above, and considering the best measures, we can state that the extended GP is able to generate better results than the standard GP either for Higgs data set or HTRU data set. Otherwise, when comparing these results with those of the state of the art, it is clear that GP, in the different explored configurations, is a competitive technique able to accomplish similar or better performance.

These preliminary experiments allow us to conclude that the solutions proposed in this work are effective and promising. However, additional experiments are needed to compare the different techniques according to a complete set of metrics such as the recall, precision and F2-score measures. Otherwise, further studies aim to better explore the different solutions and the different possibilities of combination.

6   Conclusion

With the incremental demand to analyze huge amounts of data, resulting from variant sources and generated at very high rates, researchers at different domains have studied the expansion of the existing data mining techniques to cope with the evolved nature of data. In this paper, we provide some trends to address complexity and large data size with evolutionary machine learning. Active sampling and parallelization over Spark are explained and we detailed how they can be implemented into GPs. A first experimental study to detect pulsar and Higgs Boson demonstrates how these extensions help GP to better deal with complex data. Further works aim to propose a framework implementing the different trends discussed in this paper with more comprehensive studies. In particular, implementing active sampling on top of Spark RDDs is an interesting path towards combining the discussed techniques in the same process. Another direction is studying the effect of cluster size on GP performance.

References

1. Adam-Bourdarios, C., Cowan, G., Germain, C., Guyon, I., Kegl, B., Rousseau, D.: Learning to discover: the Higgs Boson machine learning challenge (2014). http://higgsml.lal.in2p3.fr/documentation
2. Atlas, L.E., Cohn, D., Ladner, R.: Training connectionist networks with queries and selective sampling. In: Touretzky, D. (ed.) Advances in Neural Information Processing Systems 2, pp. 566–573. Morgan-Kaufmann (1990)
3. Bacardit, J., Llorà, X.: Large-scale data mining using genetics-based machine learning. Wiley Interdisc. Rev. Data Min. Knowl. Discov. **3**(1), 37–61 (2013)
4. Baldi, P., Sadowski, P., Whiteson, D.: Searching for exotic particles in high-energy physics with deep learning. Nat. Commun. **5**, 1–9 (2014)

5. Baldi, P., Sadowski, P., Whiteson, D.: Enhanced Higgs Boson to $\tau+$ τ- search with deep learning. Phys. Rev. Lett. **114**(11), 111–801 (2015)
6. Bu, Y., Howe, B., Balazinska, M., Ernst, M.D.: Haloop: efficient iterative data processing on large clusters. Proc. VLDB Endow. **3**(1–2), 285–296 (2010)
7. Chávez, F., et al.: ECJ+HADOOP: an easy way to deploy massive runs of evolutionary algorithms. In: Squillero, G., Burelli, P. (eds.) EvoApplications 2016. LNCS, vol. 9598, pp. 91–106. Springer, Cham (2016). https://doi.org/10.1007/978-3-319-31153-1_7
8. Cohn, D., Atlas, L.E., Ladner, R., Waibel, A.: Improving generalization with active learning. Mach. Learn. **15**, 201–221 (1994)
9. ATLAS Collaboration: Dataset from the ATLAS Higgs Boson machine learning challenge 2014 (2014). http://opendata.cern.ch/record/328. https://doi.org/10.7483/OPENDATA.ATLAS.ZBP2.M5T8
10. Cummins, R., O'Riordan, C.: Evolved term-weighting schemes in information retrieval: an analysis of the solution space. Artif. Intell. Rev. **26**(1–2), 35–47 (2006). https://doi.org/10.1007/s10462-007-9034-5
11. Dean, J., Ghemawat, S.: MapReduce: simplified data processing on large clusters. In: Brewer, E.A., Chen, P. (eds.) 6th Symposium on Operating System Design and Implementation (OSDI 2004), San Francisco, California, USA, 6–8 December 2004, pp. 137–150. USENIX Association (2004)
12. Fortin, F.A., De Rainville, F.M., Gardner, M.A., Parizeau, M., Gagné, C.: DEAP: evolutionary algorithms made easy. J. Mach. Learn. Res. **13**, 2171–2175 (2012)
13. Gathercole, C., Ross, P.: Dynamic training subset selection for supervised learning in Genetic Programming. In: Davidor, Y., Schwefel, H.-P., Männer, R. (eds.) PPSN 1994. LNCS, vol. 866, pp. 312–321. Springer, Heidelberg (1994). https://doi.org/10.1007/3-540-58484-6_275
14. Harding, S., Banzhaf, W.: Implementing cartesian genetic programming classifiers on graphics processing units using GPU. NET. In: Proceedings of the 13th Annual Conference Companion on Genetic and Evolutionary Computation, pp. 463–470 (2011)
15. Hmida, H., Ben Hamida, S., Borgi, A., Rukoz, M.: Hierarchical data topology based selection for large scale learning. In: Ubiquitous Intelligence & Computing, Advanced and Trusted Computing, Scalable Computing and Communications, Cloud and Big Data Computing, Internet of People, and Smart World Congress, 2016 International IEEE Conferences, pp. 1221–1226. IEEE (2016)
16. Hmida, H., Ben Hamida, S., Borgi, A., Rukoz, M.: Genetic programming over spark for Higgs Boson classification. In: Abramowicz, W., Corchuelo, R. (eds.) BIS 2019. LNBIP, vol. 353, pp. 300–312. Springer, Cham (2019). https://doi.org/10.1007/978-3-030-20485-3_23
17. Hmida, H., Hamida, S.B., Borgi, A., Rukoz, M.: Sampling methods in genetic programming learners from large datasets: a comparative study. In: Angelov, P., Manolopoulos, Y., Iliadis, L., Roy, A., Vellasco, M. (eds.) INNS 2016. AISC, vol. 529, pp. 50–60. Springer, Cham (2017). https://doi.org/10.1007/978-3-319-47898-2_6
18. Hmida, H., Hamida, S.B., Borgi, A., Rukoz, M.: Scale genetic programming for large data sets: case of Higgs Bosons classification. Procedia Comput. Sci. **126**, 302–311 (2018). The 22nd International Conference, KES-2018
19. Hunt, R., Johnston, M., Browne, W., Zhang, M.: Sampling methods in genetic programming for classification with unbalanced data. In: Li, J. (ed.) AI 2010. LNCS (LNAI), vol. 6464, pp. 273–282. Springer, Heidelberg (2010). https://doi.org/10.1007/978-3-642-17432-2_28

20. Iba, H.: Bagging, boosting, and bloating in genetic programming. In: Banzhaf, W., et al. (eds.) Proceedings of the Genetic and Evolutionary Computation Conference on GECCO-99, pp. 1053–1060. Morgan Kaufmann, San Francisco (1999)
21. Koza, J.R.: Genetic programming: on the programming of computers by means of natural selection. Stat. Comput. **4**(2), 87–112 (1994). https://doi.org/10.1007/BF00175355
22. Kuscu, I.: Genetic programming and incremental approaches to solve supervised learning problems (1996)
23. Langdon, W.B.: Graphics processing units and genetic programming: an overview. Soft Comput. **15**(8), 1657–1669 (2011)
24. Lasarczyk, C.W.G., Dittrich, P., Banzhaf, W.: Dynamic subset selection based on a fitness case topology. Evol. Comput. **12**(2), 223–242 (2004). https://doi.org/10.1162/106365604773955157
25. L'Heureux, A., Grolinger, K., ElYamany, H.F., Capretz, M.A.M.: Machine learning with big data: challenges and approaches. IEEE Access **5**, 7776–7797 (2017). https://doi.org/10.1109/ACCESS.2017.2696365
26. Lyon, R.J., Stappers, B.W., Cooper, S., Brooke, J.M., Knowles, J.D.: Fifty years of pulsar candidate selection: from simple filters to a new principled real-time classification approach. Mon. Not. R. Astron. Soc. **459**(1), 1104–1123 (2016). https://doi.org/10.1093/mnras/stw656
27. Maitre, O.: Genetic programming on GPGPU cards using EASEA. In: Tsutsui, S., Collet, P. (eds.) Massively Parallel Evolutionary Computation on GPGPUs. NCS, pp. 227–248. Springer, Heidelberg (2013). https://doi.org/10.1007/978-3-642-37959-8_11
28. Nordin, P., Banzhaf, W.: An on-line method to evolve behavior and to control a miniature robot in real time with genetic programming. Adapt. Behav. **5**(2), 107–140 (1997). https://doi.org/10.1177/105971239700500201
29. Paduraru, C., Melemciuc, M., Stefanescu, A.: A distributed implementation using apache spark of a genetic algorithm applied to test data generation. In: Genetic and Evolutionary Computation Conference, 15–19 July, Companion Material Proceedings, pp. 1857–1863. ACM (2017)
30. Peralta, D., et al.: Evolutionary feature selection for big data classification: a mapreduce approach. Math. Prob. Eng. **2015**, 11 (2015)
31. Qi, R., Wang, Z., Li, S.: A parallel genetic algorithm based on spark for pairwise test suite generation. J. Comput. Sci. Technol. **31**(2), 417–427 (2016)
32. Vanneschi, L., Poli, R.: Genetic programming - introduction, applications, theory and open issues. In: Rozenberg, G., Bäck, T., Kok, J.N. (eds.) Handbook of Natural Computing, pp. 709–739. Springer, Heidelberg (2012). https://doi.org/10.1007/978-3-540-92910-9_24
33. Wang, Y., Pan, Z., Zheng, J., Qian, L., Li, M.: A hybrid ensemble method for pulsar candidate classification. Astrophys. Space Sci. **364**, 1–13 (2019). https://doi.org/10.1007/s10509-019-3602-4
34. Zaharia, M., et al.: Resilient distributed datasets: a fault-tolerant abstraction for in-memory cluster computing. In: Proceedings of the 9th USENIX Symposium on Networked Systems Design and Implementation, NSDI 2012, 25–27 April, pp. 15–28. USENIX Association (2012)
35. Zhang, B.T., Joung, J.G.: Genetic programming with incremental data inheritance. In: Proceedings of the Genetic and Evolutionary Computation Conference, Orlando, Florida, USA, 13–17 July 1999, vol. 2, pp. 1217–1224. Morgan Kaufmann (1999). http://www.cs.bham.ac.uk/~wbl/biblio/gecco1999/GP-460.pdf

Analyzing Performances of Three Context-Aware Collaborator Recommendation Algorithms in Terms of Accuracy and Time Efficiency

Siying Li[1]([✉]) [ID], Marie-Hélène Abel[1], and Elsa Negre[2] [ID]

[1] Université de technologie de Compiègne, CNRS, Heudiasyc (Heuristics and Diagnosis of Complex Systems), CS 60 319 - 60 203, Compiègne Cedex, France
{siying.li,marie-helene.abel}@utc.fr
[2] Paris-Dauphine University, PSL Research University, CNRS UMR 7243, LAMSADE, 75016 Paris, France
elsa.negre@dauphine.fr

Abstract. Nowadays, more and more collaborative tools are available to support users' remote collaborations. Its increasing amount makes users struggle in managing and retrieving information about their collaborators during collaboration. To solve this problem, many decision support systems have been developed quickly, such as recommender systems and context-aware recommender systems. However, the performances of different algorithms in such systems are relatively unexplored. Based on our three proposed context-aware collaborator recommendation algorithms (i.e., PreF1, PoF1, and PoF2), we are interested in analyzing and evaluating their performances in terms of accuracy and time efficiency. The three algorithms all process the context of collaboration by means of ontology-based semantic similarity, but employ the similarity following two approaches respectively, to generate context-aware collaborator recommendations. In this paper, we present how to test, analyze, and evaluate the performances of the three context-aware collaborator algorithms in terms of accuracy and time efficiency.

Keywords: Context-aware recommendations · Collaborator · Collaboration context · Ontology-based semantic similarity

1 Introduction

With the development of information technology, increasingly more collaborative tools have been provided for users' remote collaborations. This makes it difficult for users to organize and manage the heterogeneous information among these tools within collaborations. Particularly, users find it difficult to retrieve information about collaborators and therefore cannot work with appropriate collaborators to advance their collaborations. To help users make decisions about

© Springer Nature Switzerland AG 2021
I. Saad et al. (Eds.): ICIKS 2021, LNBIP 425, pp. 100–115, 2021.
https://doi.org/10.1007/978-3-030-85977-0_8

collaborators, recommender systems (RS) are designed and applied [18], which can generate collaborator recommendations for them. However, such recommendations are sometimes not quite relevant, due to the missing consideration of context [3]. For instance, while generating collaborator recommendations, information of user's collaboration place and time should also be considered, which can influence users' needs for their collaborators. To incorporate context into recommendation generation processes, a new type of RS has been proposed: Context-Aware Recommender System (CARS). Despite the quick development of CARS, there is little studies on analyzing and evaluating the performance of different context-aware recommendation algorithms.

Therefore, we are interested in testing, analyzing, and evaluating the performance of our three proposed context-aware collaborator recommendation algorithms (i.e., PreF1, PoF1, and PoF2) [13] in terms of accuracy and time efficiency. The three algorithms all consider and compare the context of collaboration by means of ontology-based semantic similarity to generate context-aware collaborator recommendations for users. Their differences lie in where and how the semantic similarities are applied in recommendation generation processes. Specifically, PreF1 employs semantic similarities in the pre-processing step, following contextual pre-filtering approach [3]. As for PoF1 and PoF2, they utilize semantic similarities in post-processing step, based on contextual post-filtering approach [3]. Besides, all the three algorithms can be realized and tested in a public dataset of scientific collaborations. This allows us to evaluate their performances through three metrics: F1, mean absolute error, and execution time. These results then make it possible to analyze both advantages and disadvantages of the three algorithms in producing context-aware collaborator recommendations.

The remainder of this paper is constructed as follows. Section 2 studies context-free and context-aware recommender systems. Section 3 presents our contributions in (i) developing three context-aware collaborator recommendation algorithms (i.e., PreF1, PoF1, and PoF2) based on a collaboration context ontology and an ontology-based semantic similarity, (ii) applying a dataset to test them, (iii) analyzing and comparing their performances in terms of accuracy and time efficiency through three metrics. We then discuss the strengths and weaknesses of the three context-aware collaborator recommendation algorithms in Sect. 4. Finally, some conclusions and future work are shown in Sect. 5.

2   Related Work

This section distinguishes context-free and context-aware recommender systems (RS). Section 2.1 and 2.2 separately explore the approaches and techniques that can be used in these RSs. We then discuss their differences and relationships in Sect. 2.3.

2.1   Context-Free Recommender Systems

Context-free RSs[1] gather and utilize two types of information to provide recommendations: **user** and **item** [21]. Here, **item** is the general term to indicate the objects that 2D RS recommends to users, while **user** indicates people who will receive these recommendations [21]. For example, in a collaborator RS, an item indicates a collaborator. Particularly, an item is characterized and sorted by its ratings that indicate how a particular user liked/preferred this item [2,21]. Thus, the core task of 2D RSs is: Given an initial set of ratings that users explicitly or implicitly give for items, 2D RSs try to predict users' unknown ratings for items and decide which items to recommend [3,21]. Specifically, users' ratings for items in a 2D RS can be calculated by a function as follows [2]:

$$R_{RS} : User \times Item \rightarrow Rating \tag{1}$$

where *Rating* is a totally ordered set (e.g., non-negative integers or real numbers within a certain range).

Based on the literature review, there are four main approaches to generate 2D recommendations [1,2,21]:

1) **Content-Based filtering (CB)** approach first creates a profile for each user and a description for each item, then predicts users' unknown ratings by comparing these user profiles and item descriptions [1,2]. Usually, CB approaches apply and implement the following techniques: Vector Space Model (VSM), TF-IDF, Semantic Analysis (using ontology) [16], Naïve Bayes, Decision Trees [15], Neural Networks [19], Cosine similarity [15].

2) **Collaborative Filtering (CF)** approach predicts a particular user's unknown rating for an item based on the user's past behaviors, such as previous transactions and items' ratings [1]. The general assumption of CF approach is: if two users have same rating on one item, one of them is more likely to have the similar rating as the other on a different item [1] [2]. In CF approaches, the applicable techniques include Nearest neighbor [23], Pearson correlation [10], Cosine similarity [6], similarities in ontology [28], Matrix Factorization (MF), Latent Dirichlet Allocation (LDA), Singular Value Decomposition (SVD) [11].

3) **Knowledge-Based (KB)** approach utilizes specific domain knowledge about users' requirements and preferences to generate 2D recommendations [7]. Such approaches either determine unknown ratings by evaluating similarity metrics between the predefined cases and users' requirements [4], or estimate the extent to which an item can meet users' explicit requirements based on certain rules in a predefined knowledge base [7]. For KB approaches, techniques related to knowledge base and similarity measurement can be used and implemented, such as ontologies [25] and semantic similarity [22].

4) **Hybrid** approach combines two or more other previous approaches (i.e., CB, CF, and KB) to predict users' unknown ratings for items [2]. Thus, the techniques that can be used in the three previous approaches are also applicable in Hybrid approaches.

---

[1] Such a RS is also mentioned as **2D RS** in the rest of this paper.

2.2   Context-Aware Recommender System

As for Context-Aware Recommender System (CARS), it is constructed by incorporating context into 2D recommendation generating process, resulting in more accurate recommendations [20]. Unlike 2D RS, CARS deals with at least three types of information: **user, item** and **context**. It is even possible to construct a CARS using more types of information. For example, a context-aware movie RS [1] handled information separated into 5 dimensions: user, item, place, time and companion. Thus, the rating function of a n-dimensional CARS is [1]:

$$R_{CARS} : D_1 \times D_2 \times ... \times D_n \rightarrow Rating(n \geq 3, n \in \mathbb{N}^*) \tag{2}$$

where $D_1, D_2, ..., D_n$ represent n dimensions of information (including user, item, context, ...). When $n = 3$, it becomes: $User \times Item \times Context \rightarrow Rating$.

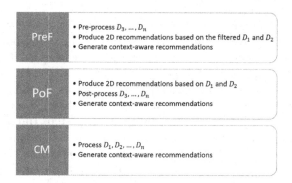

Fig. 1. Main steps of PreF, PoF, and CM [3].

Particularly, [3] proposed three approaches to incorporate the context in different steps of recommendation processes (see Fig. 1)[2], as follows:

- **Contextual pre-filtering (PreF)** approach first applies the context to filter ratings that are irrelevant to specific contexts [1]. It then employs D_1, D_2, and 2D recommendation techniques to predict unknown ratings and thus generate context-aware recommendations for users.
- **Contextual post-filtering (PoF)** approach starts with D_1, D_2, and 2D recommendation techniques, producing 2D recommendations. Afterwards, we utilize $D_3, ..., D_n$ to filter out the irrelevant 2D recommendation results or adjust their orders and thus generate context-aware recommendations [3].
- **Contextual modeling (CM)** approach produces truly multidimensional recommendations. It deals with all dimensions of information (i.e., $D_1, D_2, D_3, ..., D_n$) in each step of the whole recommendation generation process.

---

[2] Fig. 1 also illustrates how $D_1, D_2, ..., D_n$ are utilized in different methods.

2.3 Discussion

Based on the literature review of 2D RS and CARS, the techniques of CARS are still immature when compared to those of 2D RS. However, there is something in common between them. For example, techniques of some 2D recommendation approaches can directly serve in PreF and PoF approaches. This indicates that 2D recommendation approaches are important sources of inspiration for CARS, thus contributing to their development. Specifically, both PreF and PoF approaches can support any 2D recommendation approaches (i.e., CB, CF, KB, and Hybrid), but they require extra steps in generating context-aware recommendation: PreF needs to first filter out irrelevant ratings, as pre-processing step; PoF must filter out irrelevant 2D recommendation results, as post-processing step (see Fig. 1). Besides, CM approach necessitates more complicated rating functions to deal with all dimensions of information in the recommendation generation process (see Fig. 1). Consequently, any 2D recommendation approaches can not be directly applied in CM.

Moreover, the multidimensional information in CARS also causes difficulties: from the perspective of information volume, CM can not handle the same amount of users and items as the other two approaches (i.e., PreF and PoF). While CM needs to process information of $D_1 \times D_2 \times ... \times D_n$, the other two focus only on information of $D_1 \times D_2$. Meanwhile, as the dimensions of the information increase, the computational complexity of CM becomes higher and higher. However, for PreF and PoF, only their extra steps (pre-processing and post-processing) become more complicated. This indicates that, unlike PreF and PoF, CM is too costly when information volume and complexity are heavy. Thus, we concentrate on using PreF and PoF approaches to develop three context-aware collaborator recommendation algorithms (i.e., PreF1, PoF1, and PoF2) [13]. Particularly, PreF1 follows PreF approach, while PoF1 and PoF2 are built on PoF approach.

3 Contributions

This section first presents three context-aware collaborator recommendation algorithms that are developed based on a collaboration context ontology and an ontology-based semantic similarity. Then a public dataset of scientific collaborations is applied in two types of experiments to test and analyze the performances of these algorithms in terms of accuracy and time efficiency.

3.1 Context-Aware Collaborator Recommendation Algorithms

The collaboration context ontology MCC [12] employs a subject and its involved semantic 3-uples (i.e., <Subject, Predicate, Object>) to define a collaboration and its context. Specifically, a subject indicates a particular collaboration, while the related predicates and objects refer to its context which is represented by different dimensions. Thus, a collaboration c and its context are considered as a collection of semantic 3-uples with a single subject c in MCC.

Within such collections of semantic 3-uples, an ontology-based semantic similarity was proposed to measure the closeness between two collections [13], which respectively represent two collaborations (c and d) and their contexts. This semantic similarity is calculated by the following equation [13]:[3]

$$S(d, c) = \sum_{i=1}^{I} S_1^i(d, c) + \sum_{j=1}^{J} S_2^j(d, c) \tag{3}$$

where I represents the number of dimensions that contain qualitative objects; J represents the number of dimensions that contain quantitative objects; $I + J = T$, T represents the maximum number of dimensions that the context of the collaboration c can contain. Besides, $S(d, c) \in [0, 1]$. Smaller $S(d, c)$ implies greater difference between c and d.

Based on the semantic similarity $S(d, c)$ and two context-aware recommendation approaches (i.e., PreF and PoF), three context-aware collaborator recommendation algorithms can be developed: PreF1, PoF1, and PoF2. Particularly, PreF1 and PoF2 are already explained in [13]. Thus, this paper only introduces PoF1 algorithm (see Algorithm 1) and differences among the three algorithms (see Fig. 2). It first utilizes a technique of 2D recommendation approaches to predict users' unknown ratings for collaborators (line 1–3). Then, it computes ontology-based semantic similarities between the collaboration c and other collaborations d ($d \in X, d \neq c$), and a sum of semantic similarities between the collaboration c and the recent K collaborations that collaborator i ($i \neq u$) participated (line 6–13). Next, users' adjusted ratings for collaborators are computed according to predicted ratings, calculated semantic similarities, and their weights w_s, w_r (line 4–5, 14–21). Finally, the top K collaborators with higher adjusted ratings will be recommended to the user u (line 22).

PreF1 [13] first pre-processes the collaboration context by means of semantic similarity, then generates 2D collaborator recommendations, and finally produce context-aware collaborator recommendations (see Fig. 2). As for PoF1 (see Algorithm 1) and PoF2 [13], they generate 2D collaborator recommendations, then process the collaboration context using semantic similarity, and eventually produce context-aware collaborator recommendations (see Fig. 2). The differences between PoF1 and PoF2 are on how to apply semantic similarity in filtering or adjusting the order of collaborators i ($i \neq u, i \notin O^{c,Col}$).

Using the three algorithms (i.e., PreF1, PoF1, and PoF2), context-aware collaborator recommendations can be produced for users. Specifically, all these algorithms can employ the same technique of 2D recommendation approaches and our proposed ontology-based semantic similarity [13], which makes it possible to test, compare, and analyze their performances through experiments.

---

[3] The detailed equations of $S_1^i(d, c)$ and $S_2^j(d, c)$ are available in [13]. This paper do not discuss how to calculate $S_1^i(d, c)$ and $S_2^j(d, c)$.

Algorithm 1: PoF1 algorithm.

Input: The rating matrix: R,
the user: u,
the collaboration: c,
the set of members in the collaboration c: O_{Col}^c,
the number of recommendations: K,
the number of known collaborations: $|X|$,
the number of users: m,
the weight of semantic similarity in adjusted ratings: w_s,
the weight of predicted rating in adjusted ratings: w_r,
the number of dimensions that contain qualitative objects: I,
the number of dimensions that contain quantitative objects: J.
Initialize: Two lists with zeros: SC (of length $|X|$) and VU (of length m).

1   **for** $h \in \{1, 2, .., m\}$ *AND* $k \in \{1, 2, .., m\}$ **do**
2   **if** R_{hk} *is unknown* **then**
3   apply techniques of 2D recommendation approaches to predict unknown ratings R_{hk};
4   **if** $h == u$ **then**
5              $VU(k) \leftarrow 1$;

6   **for** $d \in \{1, 2, .., |X|\}$ **do**
7   **if** $d == c$ **then**
8          $S(d, c) \leftarrow 0$;
9   **else**
10          $S(d, c) \leftarrow \sum_{i=1}^{I} S_1^i(d, c) + \sum_{j=1}^{J} S_2^j(d, c)$;
11   **while** $i \in O^{c,Col}$ **do**
12   **if** *collaboration y belongs to collaborator i's rencent K collaborations* **then**
13              $SC(i).insert(S(d, c))$;

14   **for** $k \in \{1, 2, .., m\}$ **do**
15   **for** $d \in \{1, 2, .., length(SC)\}$ **do**
16   **if** $k \notin O^{c,Col}$ **then**
17              $VU(k) \leftarrow 1$;

18   **for** $k \in \{1, 2, .., m\}$ **do**
19   **if** $k \notin O^{c,Col}$ **then**
20   **if** $VU(k) == 1$ **then**
21              $\widehat{R}(k) \leftarrow w_s * sum(SC(k)) + w_r * R_{uk}$;

22   Rank $\widehat{R}$ in decreasing order and get K highest ratings;
Output: User u's adjusted ratings $\widehat{R}$, K collaborators with K highest ratings.

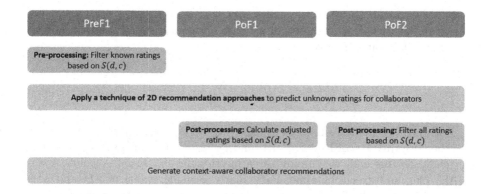

Fig. 2. Main steps in PreF1, PoF1, and PoF2.

3.2 Experiments

Dataset and Evaluation Methods. To test and evaluate the performances of PreF1, PoF1, and PoF2, we apply an academic publication dataset[4] [24]. It includes academic articles and their citation relationships until 2019-05-05, which are extracted from DBLP, MAG, and AMiner. All these articles are organized in a non-mutually exclusive hierarchy with 20 top-level domains[5]. This allows us to separate the dataset into 20 blocks. Each block contain articles in a top-level domain.

Particularly, each article in a block represents the fact that its authors have collaborated once in a top-level domain. During such a collaboration, the authors work together to write the corresponding article. This implies that an academic article can be considered as a scientific collaboration. Thus, the side information of an article belongs to the context of a scientific collaboration. Accordingly, each block of articles can be considered as a set of scientific collaborations with their contexts in the corresponding top-level domain.

Specifically, every set is composed of 1000 articles selected randomly from a block due to the different article numbers in these blocks. These articles are arbitrarily divided into two parts, representing separately training collaboration and testing collaborations. Particularly in the three context-aware collaborator recommendation algorithms (i.e., PreF1, PoF1, and PoF2), training collaborations constitute the set X, providing known information (e.g., known ratings). Meanwhile, each testing collaboration is c. For every author u in a testing collaboration c, context-aware collaborator recommendations are generated and analyzed in our experiments.

---

[4] This dataset can downloaded from https://www.aminer.org/citation.

[5] These domains are Art, Biology, Business, Chemistry, Computer science, Economics, Engineering, Environmental science, Geography, Geology, History, Materials science, Mathematics, Medicine, Philosophy, Physics, Political science, Psychology, Sociology, and Others.

Particularly, for all scientific collaborations, we have $T = 7$, $I = 1$, and $J = 6$. Therefore, based on Eq. 3, the ontology-based semantic similarity between two scientific collaborations (c and d)[6] is:

$$S(d, c) = S_1(d, c) + \sum_{j=1}^{6} S_2^j(d, c) \tag{4}$$

Within this semantic similarity $S(d, c)$, we can conduct experiments to analyze the performances of PreF1, PoF1, and PoF2 in terms of accuracy and time efficiency. To this end, we employ three metrics: F1 [8], Mean Absolute Error (MAE) [26], and execution time. Particularly, a higher value of F1[7] and/or a lower value of MAE[8] indicates more accurate recommendations. As for time efficiency, we measure the execution time[9] that each algorithm takes to generate context-aware collaborator recommendations for an author u in a testing collaboration c. Lower execution time represents higher time efficiency.

With the three metrics (i.e., F1, MAE, and execution time), two types of experiments are carried out: experiment with different semantic similarities and experiment with different 2D recommendation techniques.

Experiment with Different Semantic Similarities. In this experiment, we change the semantic similarity applied in PreF1, PoF1, and PoF2. This provides us an opportunity to investigate how much our semantic similarity [13] contributes to generating context-aware collaborator recommendations. Particularly, our semantic similarity is inspired and developed by other types of semantic similarities [13]. Thus, we need to utilize existing semantic similarities of these types in PreF1, PoF1, and PoF2. Specifically, our experiments involves 5 existing semantic similarities, including Jaccard, Dice, Tversky, TF-IDF, and IC[10] [22].

Each experiment uses a different semantic similarity in a set of scientific collaborations. To analyze their performances in the algorithms, F1, MAE and execution time are employed. All the results are shown in Fig. 3, 4, and 5.

In Fig. 3, our semantic similarity leads to the highest value of F1 when applied in PoF1 and PoF2. This indicates that our semantic similarity can enhance F1 and thus generate more accurate context-aware collaborator recommendations. But it only achieve a medium value of F1 in PreF1. These varied values of F1 imply that the enhancing effect of our semantic similarity depends on the recommendation algorithms, which is driven by the different processing steps of semantic similarities in the algorithms (see Fig. 2).

---

[6] Here, c is a testing collaboration; $|X|$ represents the number of training collaborations; $d(d \in X, d \neq c)$ is a training collaboration.

[7] The range of F1 is $[0, 1]$.

[8] The range of MAE is $[0, +\infty)$.

[9] In our experiments, execution time is counted in milliseconds.

[10] Here, IC represents $IC(c) = -\log p(c)$, where $p(c)$ is the probability of c's appearance in an ontology [22].

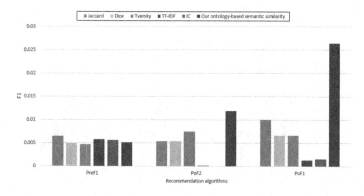

Fig. 3. F1 with different semantic similarities in a set of scientific collaborations.

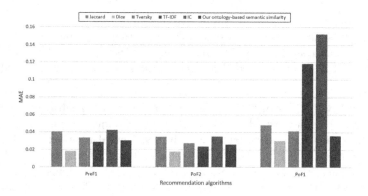

Fig. 4. MAE with different semantic similarities in a set of scientific collaborations.

Besides, when employing our semantic similarity in the algorithms, their MAE results are always lower than those of Jaccard, Tversky, and IC. However, compared to Dice and TF-IDF, our semantic similarity produces slightly higher MAE results (except TF-IDF in PoF1). Considering that our semantic similarity brings significantly better results than Dice and TF-IDF in terms of F1, it is fair to conclude that our ontology-based semantic similarity can improve the accuracy of PoF1 and PoF2. When implemented in PreF1, our ontology-based semantic similarity has an average performance of accuracy, which is acceptable.

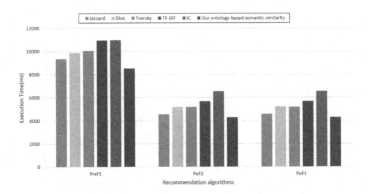

Fig. 5. Execution time with different semantic similarities in a set of scientific collaborations.

Lastly, our semantic similarity evidently achieves the shortest execution time among all semantic similarities. This indicates that the time efficiencies of PreF1, PoF1, and PoF2 are augmented through the use of our semantic similarity.

Experiment with Different 2D Recommendation Techniques. In this experiment, we employ two 2D recommendation techniques in the three context-aware collaborator recommendation algorithms: Neural network-based Collaborative Filtering (NCF) [9] and Probabilistic Matrix Factorization (PMF) [17]. Both techniques belongs to CF approach (cf. Sect. 2.1). Applying the two techniques in PreF1, PoF1, and PoF2 gives us a chance to explore whether applying our semantic similarity with different 2D recommendation techniques can influence the performances of the three algorithms. Similarly, the accuracy and time efficiency are also evaluated through the three metrics: F1, MAE and execution time. All the results are shown in Fig. 6, 7, and 8.

In summary, when either PMF or NCF is applied in PreF1, PoF2 and PoF1, all values of F1 are higher than those of PMF and NCF themselves (represented as NCF/PMF in Fig. 6, 7, and 8). This indicates that our semantic similarity can increase F1 of PreF1, PoF2 and PoF1 whatever 2D recommendation technique is used. However, F1 values with NCF are much higher than those with PMF. This implies that the enhancement of F1 driven by our semantic similarity relates to the techniques used to generate 2D collaborator recommendations. Between NCF and PMF, our semantic similarity can attain greater F1 with NCF.

Meanwhile, compared with NCF/PMF, lower MAE are also reached when applying PMF and NCF in PreF1, PoF2 and PoF1 (except PMF in PoF1). This signifies that our semantic similarity, while using either PMF or NCF in the algorithms, can reduce values of MAE. Similarly, the decreased MAE values with NCF are larger than those with PMF. This also means that the reduction of MAE is caused by our semantic similarity but linked to the techniques used to generate 2D collaborator recommendations. Both higher F1 and lower MAE

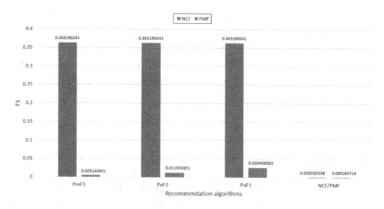

Fig. 6. F1 with different 2D recommendation techniques in a set of scientific collaborations.

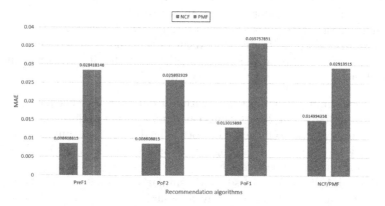

Fig. 7. MAE with different 2D recommendation techniques in a set of scientific collaborations.

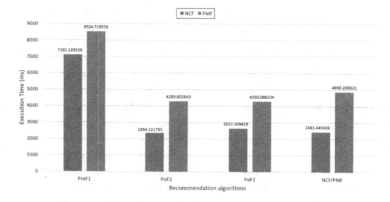

Fig. 8. Execution time with different 2D recommendation techniques in a set of scientific collaborations.

suggest that our semantic similarity improves the accuracy of PreF1, PoF2, and PoF1. Only its enhancements are more evident with NCF than with PMF.

As for execution time, the results of PoF2, PoF1, and NCF/PMF are very near. On the contrary, the execution time of PreF1 is much longer than those of other algorithms, indicating a higher time complexity of PreF1.

4   Discussion

Utilizing a dataset of scientific collaborations, we conduct two types of experiments to analyze the performance of three context-aware collaborator recommendations algorithms (i.e., PreF1, PoF1, and PoF2) from two aspects: accuracy and time efficiency. Based on the results, the following can be concluded:

- Our ontology-based semantic similarity [13] can lead to better time efficiency of PreF1, PoF1, and PoF2. It also leads to higher accuracy in PoF1 and PoF2, but shows no improvement of accuracy in PreF1.
- Calculating our ontology-based semantic similarity in PreF1, PoF1, and PoF2 can produce more accurate context-aware collaborator recommendations, regardless of the applied 2D recommendation technique. However, it positive effect on time efficiency isn't very obvious.

The above summary signifies that with our ontology-based semantic similarity, PreF1, PoF1, and PoF2 algorithms can enhance their performances, either in terms of accuracy or time efficiency, or both. The differences in improvement may be caused by the order of processing our semantic similarity in the algorithms, which deserves further investigation.

Besides, our ontology-based semantic similarity enables us to deal with the collaboration context in the context-aware collaborator recommendation generation processes. This type of context is never considered or discussed in previous studies on such recommendations. These previous studies usually focus on either user context (e.g., [14]) or item context (e.g., [27]). None of the previous studies takes into account users and items together. With the collaboration context, users and items are considered jointly in collaborations. This deepens the comprehension of both users and items in context-aware collaborator recommendations.

Also, this work has some limitations. The recommendations produced by PreF1 and PoF2 [13] may have a serendipity problem [5]: the recommended collaborators are not 'surprising' to users. This leads to insufficient diversity of recommendations: it is often the same collaborators that are recommended to users. Based on PreF1 and PoF2 [13], each collaborator recommended to the user must have involved in collaborations with high similarities. Such collaborators are obvious to facilitate users' collaborations, but not the good ones. The collaborators have not involved in the same collaborations with the user should also be included, which may result in surprising and diverse recommendations.

5   Conclusion and Future Work

In this paper, we focus on how to analyze the performances of three context-aware collaborator recommendation algorithms. To solve this issue, two types of experiments are effected in a dataset of scientific collaborations and analyzed in terms of accuracy and time efficiency.

Based on the literature review, we decided to follow PreF and PoF approaches for developing three corresponding context-aware collaborator algorithms: PreF1, PoF1, and PoF2. We then presented how the three algorithms can be used to generate context-aware collaborator recommendations based on a collaboration context ontology and an ontology-based semantic similarity. Finally, a public dataset of scientific collaborations was applied in two types of experiments to test and analyze the performances of the three algorithms through three metrics (i.e., F1, mean absolute error and execution time). The first two metrics serve to evaluate the accuracy of the three algorithms, while the last one aims to measure the time efficiency.

Our future work is to improve and enrich the experiments of the three algorithms. This may helps us to analyze why applying a same semantic similarity in the algorithms can result in different improvements.

References

1. Adomavicius, G., Sankaranarayanan, R., Sen, S., Tuzhilin, A.: Incorporating contextual information in recommender systems using a multidimensional approach. ACM Trans. Inf. Syst. (TOIS) **23**(1), 103–145 (2005)
2. Adomavicius, G., Tuzhilin, A.: Toward the next generation of recommender systems: a survey of the state-of-the-art and possible extensions. IEEE Trans. Knowl. Data Eng. **17**(6), 734–749 (2005)
3. Adomavicius, G., Tuzhilin, A.: Context-aware recommender systems. In: Ricci, F., Rokach, L., Shapira, B., Kantor, P.B. (eds.) Recommender Systems Handbook, pp. 217–253. Springer, Boston, MA (2011). https://doi.org/10.1007/978-0-387-85820-3_7
4. Bridge, D., Göker, M.H., McGinty, L., Smyth, B.: Case-based recommender systems. Knowl. Eng. Rev. **20**(3), 315–320 (2005)
5. De Gemmis, M., Lops, P., Semeraro, G., Musto, C.: An investigation on the serendipity problem in recommender systems. Inf. Process. Manag. **51**(5), 695–717 (2015)
6. Desrosiers, C., Karypis, G.: A comprehensive survey of neighborhood-based recommendation methods. In: Ricci, F., Rokach, L., Shapira, B., Kantor, P.B. (eds.) Recommender Systems Handbook, pp. 107–144. Springer, Boston (2011). https://doi.org/10.1007/978-0-387-85820-3_4
7. Felfernig, A., Friedrich, G., Jannach, D., Zanker, M.: Developing constraint-based recommenders. In: Ricci, F., Rokach, L., Shapira, B., Kantor, P.B. (eds.) Recommender Systems Handbook, pp. 187–215. Springer, Boston, MA (2011). https://doi.org/10.1007/978-0-387-85820-3_6
8. Goutte, C., Gaussier, E.: A probabilistic interpretation of precision, recall and F-score, with implication for evaluation. In: Losada, D.E., Fernández-Luna, J.M. (eds.) ECIR 2005. LNCS, vol. 3408, pp. 345–359. Springer, Heidelberg (2005). https://doi.org/10.1007/978-3-540-31865-1_25

9. He, X., Liao, L., Zhang, H., Nie, L., Hu, X., Chua, T.S.: Neural collaborative filtering. In: Proceedings of the 26th International Conference on World Wide Web, pp. 173–182 (2017)

10. Jin, R., Chai, J.Y., Si, L.: An automatic weighting scheme for collaborative filtering. In: Proceedings of the 27th Annual International ACM SIGIR Conference on Research and Development in Information Retrieval, pp. 337–344 (2004)

11. Koren, Y., Bell, R.: Advances in collaborative filtering. In: Ricci, F., Rokach, L., Shapira, B. (eds.) Recommender Systems Handbook, pp. 77–118. Springer, Boston, MA (2015). https://doi.org/10.1007/978-1-4899-7637-6_3

12. Li, S., Abel, M.H., Negre, E.: Towards a collaboration context ontology. In: 2019 IEEE 23rd International Conference on Computer Supported Cooperative Work in Design (CSCWD), pp. 93–98. IEEE (2019)

13. Li, S., Abel, M.H., Negre, E.: Ontology-based semantic similarity in generating context-aware collaborator recommendations. In: 2021 IEEE 24th International Conference on Computer Supported Cooperative Work in Design (CSCWD), pp. 751–756. IEEE (2021)

14. Liu, Z., Xie, X., Chen, L.: Context-aware academic collaborator recommendation. In: Proceedings of the 24th ACM SIGKDD International Conference on Knowledge Discovery & Data Mining, pp. 1870–1879. ACM (2018)

15. Lops, P., de Gemmis, M., Semeraro, G.: Content-based recommender systems: state of the art and trends. In: Ricci, F., Rokach, L., Shapira, B., Kantor, P.B. (eds.) Recommender Systems Handbook, pp. 73–105. Springer, Boston, MA (2011). https://doi.org/10.1007/978-0-387-85820-3_3

16. Middleton, S.E., Shadbolt, N.R., De Roure, D.C.: Ontological user profiling in recommender systems. ACM Trans. Inf. Syst. (TOIS) **22**(1), 54–88 (2004)

17. Mnih, A., Salakhutdinov, R.R.: Probabilistic matrix factorization. In: Advances in Neural Information Processing Systems, pp. 1257–1264 (2008)

18. Nunes, I., Jannach, D.: A systematic review and taxonomy of explanations in decision support and recommender systems. User Model. User-Adap. Inter. **27**(3–5), 393–444 (2017). https://doi.org/10.1007/s11257-017-9195-0

19. Van den Oord, A., Dieleman, S., Schrauwen, B.: Deep content-based music recommendation. In: Advances in Neural Information Processing Systems, pp. 2643–2651 (2013)

20. Palmisano, C., Tuzhilin, A., Gorgoglione, M.: Using context to improve predictive modeling of customers in personalization applications. IEEE Trans. Knowl. Data Eng. **20**(11), 1535–1549 (2008)

21. Ricci, F., Rokach, L., Shapira, B.: Introduction to recommender systems handbook. In: Ricci, F., Rokach, L., Shapira, B., Kantor, P.B. (eds.) Recommender Systems Handbook, pp. 1–35. Springer, Boston, MA (2011). https://doi.org/10.1007/978-0-387-85820-3_1

22. Sánchez, D., Batet, M., Isern, D., Valls, A.: Ontology-based semantic similarity: a new feature-based approach. Expert Syst. Appl. **39**(9), 7718–7728 (2012)

23. Sarwar, B., Karypis, G., Konstan, J., Riedl, J.: Item-based collaborative filtering recommendation algorithms. In: Proceedings of the 10th International Conference on World Wide Web, pp. 285–295 (2001)

24. Tang, J., Zhang, J., Yao, L., Li, J., Zhang, L., Su, Z.: ArnetMiner: extraction and mining of academic social networks. In: KDD'08, pp. 990–998 (2008)

25. Tarus, J.K., Niu, Z., Mustafa, G.: Knowledge-based recommendation: a review of ontology-based recommender systems for e-learning. Artif. Intell. Rev. **50**(1), 21–48 (2018). https://doi.org/10.1007/s10462-017-9539-5

26. Willmott, C.J., Matsuura, K.: Advantages of the mean absolute error (MAE) over the root mean square error (RMSE) in assessing average model performance. Climate Res. **30**(1), 79–82 (2005)
27. Xu, Y., Hao, J., Lau, R.Y., Ma, J., Xu, W., Zhao, D.: A personalized researcher recommendation approach in academic contexts: Combining social networks and semantic concepts analysis. In: PACIS, p. 144 (2010)
28. Zhang, Z., Gong, L., Xie, J.: Ontology-based collaborative filtering recommendation algorithm. In: Liu, D., Alippi, C., Zhao, D., Hussain, A. (eds.) BICS 2013. LNCS (LNAI), vol. 7888, pp. 172–181. Springer, Heidelberg (2013). https://doi.org/10.1007/978-3-642-38786-9_20

Recommender System for Online Teaching

Sarra Bouzayane[1]([✉])[iD] and Inès Saad[2,3][iD]

[1] Audensiel, 93 rue nationale, 92100 Billancourt, France
s.bouzayanne@audensiel.fr
[2] MIS Laboratory, University of Picardie Jules Verne, 33 rue Saint-Leu,
80039 Amiens, France
[3] Amiens Business School, 18 Place Saint-Michel, 80038 Amiens, France
ines.saad@esc-amiens.com

Abstract. This paper focuses on the support process within the online teaching environment, which is currently unsatisfactory because of the very limited size of the course trainers or teachers compared to the massive number of the enrolled learners who need support. Indeed, many of the learners can not appropriate the information they receive and must therefore be guided. Thus, in order to help these learners take advantage of the course they follow, we propose a tool to recommend to each of them an ordered list of "Leader learners" who are able to support him throughout his navigation in the online environment. The recommendation phase is based on a multicriteria decision making approach to periodically predict the set of "Leader learners". Moreover, since the learners' profiles are very heterogeneous, we recommend to each learner the leaders who are most appropriate to his profile in order to ensure a good understanding between them. The recommendation we propose is based on the demographic filtering and the Euclidean distance to identify the neighbourhood of the target learner. This method concerns only the higher-education teaching.

Keywords: Recommender system · Demographic filtering · Support process · Leader learner · Incremental prediction

1 Introduction

In the digital age, the learning processes within organizations or educational domains become increasingly mediated and learner-centred. The information exchange takes place via a digitized information system that encourages the active participation of users. The latter must enrich the system and then use it to seek the information they need. This new environment has challenged the traditional information processing methods which become insufficient or even ineffective, leading to the innovation of other technologies to cope with the massification and heterogeneity of the data. Recently, online teaching is becoming a necessity because of the health issues arising from Covid19. Several research

© Springer Nature Switzerland AG 2021
I. Saad et al. (Eds.): ICIKS 2021, LNBIP 425, pp. 116–127, 2021.
https://doi.org/10.1007/978-3-030-85977-0_9

studies have been conducted to investigate the importance of e-learning in different countries around the world [2,8,15].

In this paper, we deal with the case of information exchange via the online teaching environments which are digital information systems dedicated to online and open learning. The environments are accessible by a massive number of learners with heterogeneous profiles. A huge amount of data is deposited on the system, either by the teachers or voluntarily by the learners, in several formats (pdf, image, video, hypertext link). The data are consulted by the learners to be interpreted to information. This latter must be absorbed in order to infer knowledge. The learners periodically answer the activities proposed by the teachers, such as the automated tests and the peer assessment, using the inferred knowledge.

The majority of online courses is, however, led by a small number of teachers, usually one, that is generally unable to support all participants, which leads them to use the data exchanged via the forum whose accuracy and relevance are not always guaranteed. Hence our objective is to identify, among this massive number of learners, those who are able to share correct and immediate information with any learner in need. We call these learners "Leader learners". To do so, we propose an approach for the recommendation of a personalized list of "Leader learners" for each learner in need, taking into account their demographic data. The recommendation approach relies on a periodic prediction of the "Leader learners". This phase is based on the multicriteria decision making and aims to periodically predict the three preference ordered decision classes: Cl1 of the "At-risk learners", Cl2 of the "Struggling learners" and Cl3 of the "Leader learners". It takes into account the preferences of the teachers and the periodic variation of the learners' behaviour. A personalized list of the predicted "Leader learners" will be recommended for each "At-risk learner" or "Struggling learner" according to his profile. The recommendation is based on demographic filtering and must improve the information exchange between the "Leader learners" and the "At-risk learners" or the "Struggling learners".

The remainder of this paper is organized as follows: Sect. 2 shows the related work. Section 3 presents the "Leader learners" recommendation approach and details the periodic prediction process. Section 4 is dedicated to the experiments analysis. Section 5 concludes the work and advances some prospects.

2   Related Work

This section presents some previous works that proposed recommender systems within a context of online teaching like that of MOOCs (Massive Open Onlien Courses). In this section, we have chosen to deal with the case of MOOCs because it is a complex case study characterized by a massive number of learners who leave the course before its end. So, to avoid this problem of higher dropout rate, several recommender systems have been proposed.

Authors in [17] proposed a recommender model to provide each learner with a personalized list of discussions that satisfies his preferences. The recommendation requires three modelling: the modelling of the forum discussions content based on the words analysis, the modelling of the learner's preferences that are extracted from his discussion history and the modelling of the learner's social interactions with the pairs. Both collaborative and content-based filtering techniques are applied in order to predict the discussions that may be of interest to the learner. Authors in [11] proposed a recommender system to provide the learners with a rehabilitation resources list that is relevant for a given problem. This recommendation has to be more depth and less scaffolding than the forums interventions. The system is based on the crowdsourcing technique that requires a combination of what expert learners have published to solve a problem and what novice learners need. For each set of problems is assigned a theme and for each resource is granted a title, a link, a summary, a screenshot and a list of votes. Recommended resources can be voted on by learners. The MOOC teachers is empowered to modify, delete or enrich the resource. Onah [12] proposed an algorithm based on the collaborative filtering technique to recommend to a target learner the resources that are appropriate to his profile. Each learner in the system is asked to rate each used resource according to a scale from 1 to 5. Based on this rating, a prediction function is calculated to predict the degree of appreciation of the target learner to each resource. Finally, the authors in [9] proposed an integrated recommender module to provide each learner with a list of relevant learners who are available and ready to share their knowledge with the other learners. A target learner can send a private message or open a chat window with the recommended learner. He can also signal him as favoured or ignored. Unlike a favoured learner, an ignored learner will no longer be recommended to the concerned target learner. Learners are recommended based on their profiles and activities. The recommendation experiments showed a positive effect on the learners' participation level and the completeness of the MOOC [10].

The previous work offers assistance only to learners who participate in forums, who are usually a minority. Our proposal shares, so, the same perspective but with a generic and more rigorous system. First, it concerns all of the learners even those who did not participate in the forum discussion. Second, the list of recommended "Leader learners" is personalized according to the profile of the learner who is in need. Third, the predicted lists are periodically updated to take into account the evolution of learners' skills. Finally, the learners classification is based on the expertise of human decision-makers who are the teachers, which makes the decision more sophisticated.

3   Towards a Recommender System for a Personalized Information and Knowledge Transfer

In this section, we present the recommendation method that is based on a multi-criteria decision making approach. It consists of a periodic incremental prediction of the "At-risk learners", the "Struggling learners" and the "Leader learners". These latter will be recommended to the learners who need help in order to support them.

3.1   Method of Incremental Prediction of "Leader Learners"

This phase aims to predict the "At-risk learners", the "Struggling learners" and the "Leader learners" during the following period of the online course based on their static and dynamic data of the current Period. The definition of a period varies from one environment to another. It is defined by the teachers who derives the online course. It can be a week in case of MOOCs for example, or a session in case of a course published on Coursera or Udemy. This period can also be static that represents the entire course duration. So, from the beginning, the teacher designates the 'Leader leaders" who will still leaders throughout the course, and the same for the two other classes.

The periodic prediction process is based on the Dominance-based Rough Set Approach (DRSA) [6] and on an incremental algorithm for the updating of the approximations following the addition of a set of objects in the training set. It is composed of three steps such that the first and the second steps are performed at the end of each period P_i of the online course while the third step runs at the beginning of each period P_{i+1} of the same course with $i \in \{1..t-1\}$, where t is the course duration in periods. The first step is based on the construction of a family F_i of p criteria to characterize a learner's profile (for example the study level, the motivation to participate in this course, the score, etc.). For each criterion a preference ordered scale is fixed according to the personal point of view of the teachers (decider) [14]. For example, for the criterion "Study level", the preferences applied are: 1: Scholar student; 2: High school student; 3: PhD student; 4: Doctor. The step of constructing a family of criteria is detailed in [4]. Next, a K_i learning sample is constructed, containing a set of m reference learners. This sample must be representative for each of the three predefined decision classes:

- Cl1. The decision class of the "At-risk learners" corresponding to learners who are likely to dropout the course in the next period.
- Cl2. The decision class of the "Struggling learners" corresponding to learners who have some difficulties but still active on the online course environment and don't have the intention to leave it at least in the next period.
- Cl3. The decision class of the "Leader learners" corresponding to learners who are able to lead a team of learners by providing them with an accurate and an immediate response.

These three decision classes are increasingly preference ordered such that learners belonging to the decision class Cl3 are more preferred than those belonging to Cl2 and the latter are more preferred than those belonging to Cl1. Then, to each learner $L_{i,j} \in K_i$, such that $j \in \{1..m\}$, a vector of evaluations on the set of criteria will be assigned according to the predefined preference scale. Each evaluation vector must allow the teachers to classify the learner in one of the three decision classes Cl1, Cl2 or Cl3. However, given the free entry/exit of learners, the learning sample can not be stable from one period to another. It is therefore necessary to select for each period P_i a learning sample K_i' to be added to the learning sample K_{i-1} such that $K_i = K_{i-1} + K_i'$. To this end, the second step is based on an incremental algorithm we proposed to update the approximations in DRSA [6] following the addition of a set of learners to the learning sample. These approximations will be given as input to an induction algorithm for the inference of a set of decision rules in the form of "If ... Then ...". The last step consists in using the inferred decision rules in order to assign the potential learners, at the beginning of the period P_{i+1}, to one of the three decision classes Cl1, Cl2 or Cl3. We mean by "potential learners" those who are likely to be classified into one of the three decision classes. This phase is detailed in [4].

If the teacher manages a small team of learners, he can take over the classification phase without applying our DRSA-based prediction method. Our system is limited, in this regard, to making the correspondence between the "Leader learners" and those who are in need according to their profiles.

3.2 Recommendation Process Based on the "Leader Learners" Prediction

The objective of the recommender system is to provide each "At-risk learner" or "Struggling learner" with an ordered list of "Leader learners" who are able to support him during his participation in the online course. The recommendation process is based on three steps: first, the learner's profile modelling, second the learner's neighbourhood identification and finally the recommendation list prediction.

Learner's Profile Modelling. The user's profile modelling is based on two key concepts that are the representation model and the information to consider. In this work, we adopt the vector representation. Moreover, the information to be included in the representation model is entered manually by the learner upon the registration. It must satisfy the purpose of the recommendation, which in our case is the mutual understanding between the information transmitter (the "Leader learner") and the receiver (the target learner). In other words, this information must represent the factors inhibiting the process of knowledge transfer such as the language, the field of study and the geographical distance.

- Language: the language has been proven by several research works as a powerful obstacle to the process of knowledge transfer. In the knowledge management field, [16] proved that the language impacts, both the ability to transfer knowledge by the transmitter and also the ability to absorb this knowledge by the receiver. In the context of e-learning environments, [3] has proved that language is a strong barrier to the understanding and the completeness of the course.
- The field of study: the shared field of study allows the actors to have a similar scientific and technical language. According to [7], people who share the same culture may have similar patterns of interpretation that allow them to give the same meaning to a codified knowledge. Also, [5] highlighted the impact of the cultural distance on the quality of the knowledge transfer process. He

proves that people from the same culture can understand each other better than the other people.

- Geographical distance: compared to the face-to-face interaction, remote interaction puts a lot of disadvantages especially when it concerns the know-how transfer. According to [5], geographic distance is an inhibitor of the knowledge transfer process. [1] found that the smaller the distance between the transmitter and the receiver, the higher the efficiency of knowledge transfer.

This information is to be considered in order to calculate the similarity between the "Leader learner" and the target learners. It is preferable to minimize the linguistic, the cultural and the geographical distance between the knowledge transmitter and the knowledge receiver.

Learner's Neighbourhood Identification. The neighbourhood of a target learner is the set of learners who are closer to him considering their language, their field of study, their country and their city. We are thus faced with a problem of distance minimization using the Euclidean distance.

The Euclidean distance has a lower limit of 0 indicating a perfect correspondence with no proportional upper limit. The vector representations of the profiles of two learners x and y, respectively, are $(x_1, x_2, \ldots, x_k, \ldots, x_n)$ et $(y_1, y_2, \ldots, y_k, \ldots, y_n)$. The Euclidean distance between the two profiles is calculated as shown in Eq. (1):

$$d(x, y) = \sqrt{\sum_{i=1}^{n} (x_i - y_i)^2} \tag{1}$$

In our case we consider only four attributes (n = 4) which are the language, the field of study, the country and the city. For example, considering two learners X and Y characterized as follows: X = (French, Computer science, France, Paris) and Y = (French, Computer science, Belgium, Brussels). The Euclidean distance between X and Y is calculated as follows: $d(X, Y) = \sqrt{0^2 + 0^2 + 1^2 + 1^2} = \sqrt{2}$.

Prediction Based on Demographic Filtering. The demographic filtering is based on the ratings made by the demographic neighbourhood of the target user. The recommendation, in our case, is the "Leader learners" who have previously been rated and appreciated by the target learner's neighbourhood.

In order to recommend to a target learner the list of "Leader learners" the more appropriate to his profile, we must predict the rate of appreciation $\hat{r}_{c,l}$ of a target learner c for a "Leader learner" l, using the ratings given by his neighbourhood for this same "Leader learner".

$$\hat{r}_{c,l} = \frac{\sum_{(v \in V_l(c))} w_{c,v} r_{v,l}}{\sum_{(v \in V_l(c))} |w_{c,v}|} \tag{2}$$

In Eq. (2), the variables c, l, v denote respectively the target learner, the "Leader learner" and the neighbour. The set $v_l(c)$ is the neighbourhood of the target learner having already rated the "Leader learner" l. The variable $w_{c,v}$ reflects the weight of the neighbour, calculated by its degree of similarity with the target learner. The rate $r_{v,l}$ is the evaluation given by the neighbour v to the "Leader learner" l.

The denominator of Eq. (2) has been added for a standardization objective to avoid the case where the sum of the weights exceeds the value 1 which can give a predicted value out of range. Similarly, the weight was used in absolute value to deal with the case of negative evaluations that could lead to unacceptable values. More details on this measure exist in [13].

3.3   Simplifying Assumptions

In order to take into account the possible particular cases and to manage the situations of conflicts, we applied some simplifying assumptions:

– In order to cope with the high number of the "At-risk learners" and the "Struggling learners" compared to the number of the online "Leader learners", we limit the size of a recommended list to three. Also, to respect the human capacity of a "Leader learner" we propose to him a maximum of three "At-risk learners" or "Struggling learners" at a time.
– A "Leader learner" evaluated as "irrelevant" by a target learner will no longer be recommended to him, even if he has been assessed as relevant by his neighbourhood. Similarly, a "Leader learner" appreciated by a target learner will automatically be recommended on the header line of the list provided that he is online and available, thus exchanging with less than three learners.
– In case of conflict between an "At-risk learner" and a "Struggling learner", we give priority to the struggling one considering that he is more motivated to complete the course. Since an "At-risk learner" is more likely to drop out the course, it is more profitable to accompany the "Struggling learner".
– In the case of a new "Leader learner" who is not yet evaluated or the case of lack of available "Leader learners", the system completes the list to be recommended to the target learner by the online and available "Leader learners" from his Neighbourhood. In this case, the system recommends to the target learner the "Leader learners" who are similar to him instead of the "Leader learners" appreciated by the learners who are similar to him in order to face the cold start problem.

The "Leader learners" recommendation algorithm must consider these simplifying assumptions. The 3-Top online and available "Leader learners" with the highest $\hat{r}_{c,l}$ value will be recommended and displayed on the personal page of the target learner.

3.4   Functional Architecture of the Interactive Decision Making System

The functional architecture of the interactive decision making system defines the operating principle of the system. It is described in Fig. 1. The operation

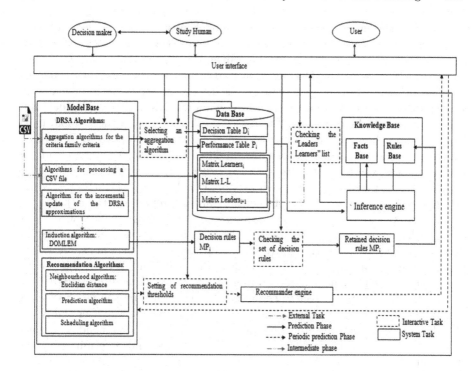

Fig. 1. Functional Architecture of the interactive decision making system

of the system starts by cleaning the CSV (Comma-Separated Values) file using the processing algorithms dedicated to this type of files. In our case, this file contains the learners' static and dynamic data which are generated by the trace analysis algorithms implemented in the online course system. The data will be stored in the matrix $Learners_i$, which gathers all the learners' data during the $Period_i$ of the course, and the matrix L-L which gathers the ratings given by the learners to each of the "Leader learners" from the beginning of the course. The CSV file is also used by the study human and the decision maker in order to collectively construct the family of criteria to characterize a learner's profile and behaviour. The study human and the decision maker have to validate the appropriate algorithm in order to aggregate the criteria to be retained and to control their redundancy. At the end of each $Period_i$ of the course, these data will be stored in the decision table D_i. This table also contains the evaluation values of each learner on the set of criteria. These values result from the processing algorithm of the CSV file of the $Period_i$. The decision table D_i contains, moreover, the assignment of each learner in one of the predefined decision classes. The decision of assignment is made by the decision maker. Next, an incremental algorithm for the update of the approximations in DRSA is applied on the decision table D_i of the $Period_i$. These approximations are given as input to the induction algorithm DOMLEM in order to infer a preference model PM_i resulting in a set of decision rules. The preference model must be checked by the study human and the decision maker to validate it or to select a consistent sub-set. At the beginning of each $Period_{i+1}$ of the course, the performance table P_i is filled using the CSV file of the $Period_i$. The learners in the performance table are called "Potential learners". These are the learners who are likely to be classified in one of the predefined decision classes. The information contained in this table will be transformed into $Facts_i$. The decision rules selected in the $Period_i$ must be applied on these $Facts_i$ to predict the "Leader learners" of the $Period_{i+1}$. The list of "Leader learners" must be validated by the study human and the decision maker who can capitalize on their experience to validate it or to define a final sub list. The leaders' data will be stored in the matrix $Leaders_{i+1}$, which is updated in real time. The list of "Leader learners" represents the items in the recommender system. The "At-risk learners" and the "Struggling learners", once connected, represent the target users of the recommendation. The demographic data about the users (learners) are stored in the matrix $Learners_i$. A neigh-bourhood algorithm based on the Euclidean distance is applied on this matrix to identify the neighbourhood of each user. A prediction algorithm to predict the user's appreciation about each "Leader learner" is applied on the *matrix L-L*. A scheduling algorithm is finally applied in order to display to the user the N-top "Leader learners" appropriate to his profile. These three recommendation algo-rithms are based on thresholds that must be discussed and determined by the study human and the decision maker such as the number of "Leader learners" to be recommended.

4   Experiments and Results

In this section we evaluate the proposed recommender system on two aspects: the run time performance and the space coverage. All algorithms in this paper are coded by Java and were run on a personal computer with Windows 7, Intel (R) $Core^{TM}$ i3-3110M CPU @ 2.4 GHz and 4.0 GB memory.

4.1   Run Time Performance of the Recommendation Algorithm

Figure 2 shows the results of simulations performed according to the variation of the number of learners and also the number of ratings recorded in the database. We find that the proposed algorithm is sensitive to the two factors. The recom-mender algorithm is faster when fewer learners are enrolled and fewer evaluations are given. This seems logical because demographic filtering processes the demo-graphic data of all enrolled learners as well as the assessments they submit.

4.2   Evaluation of the Item Space Coverage

It is also important to evaluate the coverage of the item space ("Leader learn-ers") of the proposed system. This coverage refers to the proportion of "Leader

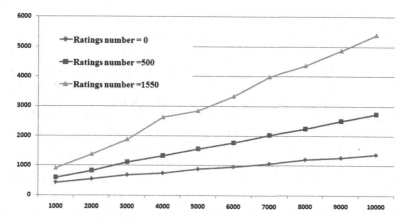

Fig. 2. Execution time of the recommendation algorithm: study of the algorithm sensitivity in relation to the number of the learners and that of the stored ratings

learners" recommended by the system. It represents the percentage of the recommended "Leader learners" in relation to the total number of "Leader learners". In our case the space item coverage depends on the number of "Leader learners" available and also on the number of "At-risk learners" or "Struggling learners" that we plan to help.

Figure 3 shows the results of the simulations performed on separate sets of data by modifying the size of the recommendation target set. The "Leader learners" are ordered in increasing order of frequency.

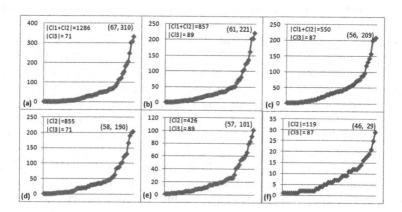

Fig. 3. Coverage of the space item and diversity of proposed recommendations (X axis: Leader learner's identifier recommended; Y axis: Number of recommendations of a leading learner).

In the upper curves ((a), (b) and (c)), the recommendation concerns the "At-risk learners" classified in the decision class Cl1 and the "Struggling learners" classified in the decision class Cl2. However, in the lower curves ((d), (e) and (f)) the recommendation concerns only the "Struggling learners". We find that the coverage on the item space decreases by decreasing the target set size of the recommendation. Indeed, if we consider the ratio between the maximal number of recommended "Leader learners" and the total number of the "Leader learners" we found: $\frac{67}{71} = 0.94$, $\frac{61}{89} = 0.68$ and $\frac{56}{87} = 0.64$ respectively for curves (a), (b) and (c), $\frac{58}{71} = 0.81$, $\frac{57}{89} = 0.64$ and $\frac{46}{87} = 0.52$ respectively for the curves (d), (e) and (f). This is due to the heterogeneity of the learners profiles, which decreases according to their number and also to the marker we have imposed on the size of the recommendation list (3 "Leader learners" only can be recommended for each target learner) which prevented find the demographic pairs for some "Leader learners" and therefore we get smaller coverage. In this case, a "Leader learner" will be recommended several times which influences the diversity of the recommendation. However, we should note that in all of cases, more than half of the "Leader learners" were recommended.

In other hand, we used a set of real data coming from a French MOOC in order to asses the quality of the recommended "Leader learners" [4]. Experiments proved a satisfactory F-measure[1] that reaches 0.67. This said, the majority of the recommended learners were truly leaders.

5   Conclusion

In this paper we proposed a recommender system that provides each "At-risk learner" or "Struggling learner" participating in the online course with a list of "Leader learners" appropriate to his profile. The "Leader learner" has the role to support the target learner throughout his learning process by providing him with the accurate and immediate information he needs. Our approach is composed of two phases: a periodic incremental prediction phase of the "Leader learners" based on a multicriteria decision making approach and a recommendation phase based on the demographic filtering and the Euclidean distance measurement. The proposed recommender system is a widget integrated in the personal page of an "At-risk learner" or a "Struggling learner" containing a personalized list of three "Leader learners". Simulations conducted showed a satisfactory results considering the coverage (from 0.52% to 0.94%) and the execution time of our recommender system. The quality of the recommended "Leader learners" was previously tested on real data from a French MOOC and proved a satisfactory F-measure that reaches 0.67. In our future work, we intend to experiment the proposed recommender system on other online teaching environments and to compare it with other previous works.

---

[1] F-measure is a measure of a test's accuracy. It is calculated from the precision and recall of the test, where the precision is the number of true positive results divided by the number of all positive results, including those not identified correctly, and the recall is the number of true positive results divided by the number of all samples that should have been identified as positive.

References

1. Ambos, T.C., Ambos, B.: The impact of distance on knowledge transfer effectiveness in multinational corporations. J. Int. Manag. **15**(1), 1–14 (2009)
2. Bao, W.: COVID-19 and online teaching in higher education. Human Behav. Emerg. Technol. **2**(2), 113–115 (2020)
3. Barak, M.: The same MOOC delivered in two languages: examining knowledge construction and motivation to learn. In: Proceedings of the EMOOCS, pp. 217–223 (2015)
4. Bouzayane, S., Saad, I.: A multicriteria approach based on rough set theory for the incremental Periodic prediction. Eur. J. Oper. Res. **286**(1), 282–298 (2020)
5. Gooderham, P.N.: Enhancing knowledge transfer in multinational corporations: a dynamic capabilities driven model. Knowl. Manag. Res. Pract. **5**(1), 34–43 (2007)
6. Greco, S., Matarazzo, B., Slowinski, R.: Rough sets theory for multicriteria decision analysis. Eur. J. Oper. Res. **129**(1), 1–47 (2001)
7. Grundstein, M.: GAMETH®: a constructivist and learning approach to identify and locate crucial knowledge. Int. J. Knowl. Learn. **5**(3–4), 289–305 (2009)
8. Henaku, E.A.: COVID-19 online learning experience of college students: the case of Ghana. Int. J. Multidiscip. Sci. Adv. Technol. **1**(2), 54–62 (2020)
9. Labarthe, H., Bachelet, R., Bouchet, F., Yacef, K.: Increasing MOOC completion rates through social interactions: a recommendation system. Research Track, p. 471 (2016)
10. Labarthe, H., Bouchet, F., Bachelet, R., Yacef, K.: Does a peer recommender foster students' engagement in MOOCs? International Educational Data Mining Society (2016)
11. Li, S. W. D., Mitros, P. : Learnersourced recommendations for remediation. In 2015 IEEE 15th International Conference on Advanced Learning Technologies, pp. 411–412. IEEE (2015)
12. Onah, D. F. O., Sinclair, J. E.: Collaborative filtering recommendation system: a framework in massive open online courses. In: INTED2015 Proceedings, pp. 1249–1257 (2015)
13. Ricci, F., Rokach, L., Shapira, B.: Introduction to recommender systems handbook. In: Ricci, F., Rokach, L., Shapira, B., Kantor, P.B. (eds.) Recommender Systems Handbook, pp. 1–35. Springer, Boston, MA (2011). https://doi.org/10.1007/978-0-387-85820-3_1
14. Roy, B., Mousseau, V.: A theoretical framework for analysing the notion of relative importance of criteria. J. Multi-Criteria Decisi. Anal. **5**(2), 145–159 (1996)
15. Van der Spoel, I., Noroozi, O., Schuurink, E., van Ginkel, S.: Teachers' online teaching expectations and experiences during the Covid19-pandemic in the Netherlands. Eur. J. Teach. Educ. **43**(4), 623–638 (2020)
16. Welch, D.E., Welch, L.S.: The importance of language in international knowledge transfer. Manag. Int. Rev. **48**(3), 339–360 (2008). https://doi.org/10.1007/s11575-008-0019-7
17. Yang, D., Piergallini, M., Howley, I., Rose, C.: Forum thread recommendation for massive open online courses. In: Educational Data Mining (2014)

Security, Artificial Intelligence, and Information Systems

Artificial Intelligence & Cybersecurity: A Preliminary Study of Automated Pentesting with Offensive Artificial Intelligence

Marin François[1]($\boxtimes$) (ID), Pierre-Emmanuel Arduin[2]($\boxtimes$), and Myriam Merad[3]($\boxtimes$)

[1] Université Paris-Dauphine, PSL, Paris, France
marin.francois@dauphine.eu
[2] Université Paris-Dauphine, PSL, CNRS, DRM, Paris, France
pierre-emmanuel.arduin@dauphine.psl.eu
[3] Université Paris-Dauphine, PSL, CNRS, LAMSADE, Paris, France
myriam.merad@dauphine.psl.eu

Abstract. In this paper, we seek to define an experimental framework for the application of a new industrialization method for penetration testing. This work-in-progress research is placed in a particular business context: that of a company with an extensive and decentralized information system. The objective of this research is to give companies the tools to develop a penetration test task force capable of testing any system in a fully automated way and to form proper communication channel and support for risk assessment reporting. It is based on the use of artificial intelligence to make the penetration test autonomous. This research considers the conduct of penetration tests both through their technical issues and through the managerial issues specific to a decentralized information system.

Keywords: Information systems · Cybersecurity · Penetration-testing · Machine learning

1 Introduction

The implementation of an Information Security Management System (ISMS) is the main purpose of the ISO/IEC 27001:2018 [8] standard, an essential reference for any company wishing to improve the security of its information system and to capitalize on this security. The implementation of an ISMS provides reasonable assurance of an organization's ability to provide effective risk management. On the one hand, it concerns the company's Information Technology (IT) needs in an increasingly active cyber-space, but on the other hand, a ISO27001 certification is a guarantee for the company's customers. However, the sustainability of the ISMS is intrinsically dependent on the frequency of testing of this system [3] and the controls it is subject to [3]. Indeed, as the information system is for many companies evolutive (configuration changes, addition and removal of components, updating of services and operating systems, but also evolution of attacker behaviors and malicious tools), we argue that a simple spot check of the security controls

© Springer Nature Switzerland AG 2021
I. Saad et al. (Eds.): ICIKS 2021, LNBIP 425, pp. 131–138, 2021.
https://doi.org/10.1007/978-3-030-85977-0_10

and systems only allows to have an estimation of the level of risk on a precise time window. As soon as there is a change in the environment, either of the Information System (IS) or the attackers, the risk also changes.

How do you give a company with an extended information system the ability to conduct automated penetration tests on its entire perimeter? How can we include in the framework specific tools for risk analysis in a decentralized information system?

In this research, we propose new tools for cyber risk management through the practice of penetration testing. On one hand, the implementation of an "understandable by anybody" fine granularity reporting system for the feedback of information and visibility on the vulnerabilities of a large perimeter in the context of a multinational company (with the implications linked to a decentralized and extended information system). In the other hand, the deployment of a large-scale testing solution allowing to conduct automatic and flexible penetration tests. These tests include an analytical dimension with that would provide an added value during the decision-making process related to the securing of the IT infrastructures. There is therefore a methodological and strategic dimension here, relating to risk management, but also an IT and purely technical dimension relating to security engineering.

Firstly, we will look at the state of the art of pentesting and the emerging need for new tools. Then we look at the definition of an extended information system, its characteristics, and the implications of this strategy on the work of the pentester. The specificities of the environment in which the software will be deployed also imply questioning the collection of the information necessary to define the software model. Finally, it is important to consider the human aspects of the information system, and in particular the impact of responsibility and hierarchy on the flow of information.

2   Background

In this section, we will focus on three concepts: the practice of penetration testing, the components of an extended decentralized information system and the issue of uniform risk quantification. These three notions are essential as they are the building blocks of our research framework.

2.1   Penetration Testing

A penetration test or pentest is a method of evaluating (auditing) the security of an information system or computer network; it is performed by a pentester. These tests can be conducted under three levels of preparation: "black box" (no connection information prior to the attacks), "grey-box" (brief information on the infrastructure, configurations, and systems) and "white box" (technical documentation, connection information and advanced knowledge of the systems, configurations, and infrastructure tested). Sometimes referred to as "red-teaming" as opposed to "blue-teaming" defensive security work, penetration testing is a form of offensive security that allows a person to put themselves in the shoes of an attacker on a network in an attempt to find and exploit vulnerabilities. Penetration testing requires a strong understanding of computer systems and architectures, as well as the ability to adapt to each new situation. Corrective measures can then be taken to secure the infrastructure [12]. These tests can be conducted according to a predefined attack scenario (for example, an attack that has been particularly devastating in the past). They can also be conducted in a more open- ended manner: we know where the test starts, but we don't know where it ends. The objective of the "white hat", or ethical hacker [11], is to compromise a predefined perimeter: this can be done by recovering confidential information, stealing passwords, etc. Once the test is completed, the managers of the tested perimeter receive a risk analysis and a remediation action plan weighted by the white-hats findings.

Penetration tests are most of the time conducted according to the same roadmap: discovery, scanning, exploitation, post-exploitation, information classification, remediation proposal, counter-audit [2].

If a company were to develop a centralized task force as presented, then it would have to develop a huge testing capacity in terms of volume. It would also have to capitalize on an extremely large knowledge base [4]. Automated penetration testing methods are now beginning to emerge, with some soft- ware even including attack strategies based on machine learning or reinforcement learning [9]. Machine learning has the advantage of being faster to implement, as the model is faster to train depending on the type of optimization chosen [13], but it requires the inclusion of a large amount of data for training. Reinforcement learning, on the other hand, has the advantage to offer the possibility of being more restricted in training data because the evolution of the intelligent agent is primarily based on the context [7]. However, reinforcement learning has the disadvantage to be longer to implement. Based on the use of time-series, these two models are indeed suitable for modeling the behavior of an attacker on a network. The measured variable can be, for example, an executed bash command. It is now necessary to ask ourselves in what context, in what environment this pentesting system will be integrated.

2.2 Extended Information System

In the interests of efficiency, some global companies choose to abandon a vertical organization in favor of a horizontal and decentralized one [16]. Structured in the form of poles (e.g. by continent/subcontinent), this organizational scheme allows for faster decision-making and a more flexible organization [1]. This international management strategy is based on the assumption that in some cases local experts are better able to catalyze expansion through their local knowledge than centralized decision-makers [16]. Whereas in a traditional horizontal scheme, information flows directly from point A to point B, in these structures, information takes the form of clusters before flowing sequentially to higher decision-makers. Although this method of management in an international environment has the advantages mentioned above, it can be a hindrance to the application of good risk management: rapid development of shadow IT [6], lack of visibility on the controls applied in the various business units, but above all, the difficulty to conduct real-time and large-scale penetration tests.

Thus, this extended information system is difficult to test on a large scale. Centralized risk managers and decision-makers must call upon local cluster managers, who call upon local security expertise.

Conducting penetration tests is costly and time consuming when done manually. Often outsourced, it leads to significant costs and no real gain in visibility on the tested perimeters since they are managed locally. The main issue here is the implementation of an action and communication chain (dedicated to penetration testing) based on automated and centralized management, in a decentralized and extended international environment, operating on sequential management in geographical clusters. One possible solution to this problem of scale and reporting to decision-makers could be the creation of a centralized penetration testing task force that would cover the entire scope of the company, as notably tested by VINCI Energies.

3   Towards a Unified Risk Grid

When operating in an extended decentralized information system, one of the main problems in the feedback of information between different actors is the interpretation in relation to common repositories. Regarding the risk assessment, the understanding of the risk following a penetration test is often divided into two categories: on the one hand "technical" people with IT skills, and on the other management, i.e. "business-oriented" people with skills in business management. Where the former will need technical information on computer components and vulnerabilities, the latter will want a quantified interpretation (from a financial point of view) of the risk. The implementation of an information transmission chain that can suit all the stakeholders involves the use of metrics that can be understood by all. Since the financial quantification of intangible risks constitutes the primary profession of insurers, we believe that it would be possible to borrow quantification methods from these same insurers in order to read cyber risk under a financial grid. This is particularly what the FAIR method could allow [17]. This unified reading grid could be a partial solution to the problem of transmitting information within a heterogeneous, decentralized, extended group.

The principles of risk analysis and treatment are now well known and mastered by security professionals. The risk approach is the foundation of security programs. ISO27005, NIST 800-30, CRAAM, all of these methods make it possible to formalize risk levels (threats, vulnerability, impacts, likelihood and residual risk after treatment) but not to quantify a risk financially. The FAIR (Factor Analysis of Information Risk) method follows a scientific approach to calculating risk, defined as the likely frequency (taking into account frequency of occurrence and vulnerability) and the likely magnitude of loss. The introduction of the notion of probability of action measures the motivation of the attacker, his perception of the level of effort and of his own risk-taking (an important notion for cases of internal fraud, the more the potential fraudster has the perception of impunity, the greater the probability that he will take action). Vulnerability is measured as a percentage representing the probability that an action will result in a loss. Impact assessment takes into account direct losses (e.g. loss of revenue) and indirect losses (e.g. financial penalties, decrease in the company's market value or damage to reputation). Statistical models can be based on the Monte Carlo method [18]. FAIR is a method that not only defines terms and concepts, but also specify how to carry out each step: how to measure the different criteria and on the basis of which data, the analysis of cyber risk scenarios and the interpretation of the results. Thus, this quantification method could be used to quantify the risk in a uniform way, based on the probabilities of cyber events occurrence.

4   Research Proposal

Based on the Design Science Research paradigm, we hope to be able to propose a development practice adapted to our field environment to achieve the develop a decision support tool for risk management. In this section we will discuss the technical requirements and development limits of our methodology and decision support tool.

4.1   Development of a Methodology and a Decision Support Tool

To meet the technical and knowledge-oriented needs of the centralized pentesting task force, managers have to look for and develop new testing methods. Automation could be a way to increase the number of penetration tests conducted within organizations. Last research results show the potential of penetration test automation [5, 14] using machine learning/reinforcement learning models. These prototypes of automated systems, generally developed in Python and shellcode, call upon the tools generally used by pentesters. For machine learning models, once the recognition and scanning steps have been completed, an attack graph can be generated based on the existence of exploits and conditions necessary to conduct the attack (e.g. the version of a known vulnerable OS and the availability of an exploit). For reinforcement learning models, it is a matter of giving a test environment to the intelligent agent (for example, a typical architecture, with its vulnerabilities and blocking mechanisms). We would then place a reward function corresponding to the execution of a code allowing the recovery of a "flag" in the manner of the "Capture the Flag" games popular with hackers.

4.2   Design Science Research Framework

We would first identify the specificities of cyber risks, compared to other types of risks, conduct a critical analysis of the methods and tools developed for the analysis and management of cyber risk, and then enhance the development of a tool and methodology for cyber risk management. This methodology and tool should consist of four parts: a serious game part (attack & defense simulation) based on the technical elements we discussed in the upper section, an analysis of system vulnerabilities (within the company) and analysis of the attack strategy (known-realized or described in the literature), a third part on the development of several modules / strategy for cyber risk management, and finally, a part on cost/benefits evaluation of these strategies (through our unified risk grid). Finally, we would conduct an application of the methodology and tool on a case study.

4.3   Limitations

There are data resources available for free in cybersecurity (Virus Total, Kaspersky, Kaggle), but not-enough for the precision results and level of accuracy we expect if using Machine Learning. Indeed, if we want to be able to define a training environment for an artificial intelligence model, we need a significant amount of data specific to the company's environment. However, some learning models, such as reinforcement learning for example, require less training data and only an evolutionary context for the agent. This could be a way out for our development [9]. The use of data related to past attacks, public data related to component vulnerabilities as well as ad-hoc data specific to the tested perimeter could provide a risk analysis solution at a fine granularity and on many IT components that the autonomous system would encounter.

When a computed solution is found to conduct the tests autonomously on a large scale, then the task force could be set up, logically speaking. From a purely architectural point of view, a solution that could be interesting for the orchestration of this tool could be the use of containers. Because the container-based architecture allows simpler deployment and better management of data chains [10]. Once a dataset is structured, it would be possible to look for: (1) an efficient penetration test conduct model and (2) an efficient risk analysis model according to the elements encountered on the tested perimeters. The former ensures the classification of information, while the later ensures the estimation of the level of risk associated with a configuration or a computer component. However, if we tried to see some possible approaches for the first one, the second question remains open: how can risk analysis be automated at a granularity fine enough to provide value? Indeed, risk analysis is not only based on attack results but also on informal elements and knowledge acquired over time [15]. On this second question, we call upon the use of FAIR analysis. With the implementation of the solution we propose, the company's ability to test large areas without requiring a large number of agents will increase. In turn, it would gain visibility into its actual day-to-day risk level, without relying on sometimes inefficient reporting channels. However, the success of this method depends on our ability to (1) develop a relevant attack model on heterogeneous perimeters, (2) design a clear development road map for Machine Learning or Reinforcement Learning development and (3) define relevant and understandable Risk-assessment metrics.

However, we can question the technical feasibility of a program being able to test any system. Moreover, if this system is feasible, the non-intrusiveness / low-intrusiveness characteristic of this system would be essential. We can also wonder about the ethical and legal implications of such a system.

5   Conclusions and Perspectives

In this article, we have defined an experimental framework aiming to answer to the following problems: (1) how to give a decentralized company with a large information system the capacity to conduct a large number of penetration tests? and (2) how to gain visibility on the real risk level of its infrastructures via comprehensible unified metrics?

The next stage of our work will now focus on the state of the art in this area, and with such a solid basis for enriching our program, we will be able to define the technical characteristics of the penetration tests automation solution. Then, we will develop the solution based on the research of an efficient artificial intelligence model with the data provided by our industrial partners. Once this software module is available, we will hopefully be able to work on proposing a homogeneous reading grid of the intrusion results.

We have considered the constraints related to the structure of decentralized and extended information systems. We also discussed the possibility of automating penetration testing tools, as well as the theory of developing an intelligent agent based on reinforcement learning to conduct these tests. Finally, we have discussed the interest of a unified risk grid, based on a finer granularity. This grid can be obtained by developing quantitative methods like the FAIR method. Considering these elements, we are now working on the development of a methodology and a decision support tool for cyber risk management. This tool and methodology include a simulation component, based on the reproduction of the company's IT environment, in which an agent will evolve to simulate attack and defense. This tool will then allow us to obtain information in order to quantify the risk in a more precise way and to evaluate the costs and benefits of the various defense strategies. However, we will encounter several difficulties: mainly the lack of data to train the intelligent agent (why we chose reinforcement learning), but also the technical complexity of the information system we have to reproduce.

References

1. Aoki, M.: Orizontal vs. Vertical information structure of the firm. Am. Econ. Rev. **76**(5), 971–983 (1986)
2. Bertoglio, D.D.: An overview of open issues on penetration test. J. Braz. Comput. Soc. **23**(1), 1–16 (2017)
3. Bialas, A.: A UML approach in the ISMS implementation. In: Dowland, P., Furnell, S., Thuraisingham, B., Wang, X.S. (eds.) Security Management, Integrity, and Internal Control in Information Systems, pp. 285–297. Springer, Heidelberg (2004). https://doi.org/10.1007/0-387-31167-X_18
4. Sarraute, C.: Using AI techniques to improve Pentesting Automation. Hackito Ergo Sum (HES), Paris, France (2010)
5. Chowdhary, A., Huang, D., Mahendran, J. S., Romo, D., Deng, Y., Sabur, A.: Autonomous security analysis and penetration testing. In: 16th International Conference on Mobility, Sensing and Networking (MSN), pp. 508–515 (2020)
6. Rentrop, C., Zimmermann, S.: Shadow IT: management and control of unofficial IT. In: 6th International Conference on Digital Society (ICDS), pp. 98–102 (2012)
7. Dayan, P.: Reinforcement Learning. Wiley, Hoboken (2002)
8. ISO/IEC 27001: Information technology—security techniques—information security management systems—requirements. Technical Report ISO/IEC 27001, ISO/IEC, Geneva (2018)
9. Schwartz, J., Kurniawati, H.: Autonomous penetration testing using Reinforcement Learning. Cornell University Computer Science (2019)
10. Bravo Ferreira, J., Cello, M., Iglesias, J.O.: More sharing, more benefits? A study of library sharing in container-based infrastructures. In: Rivera, F.F., Pena, T.F., Cabaleiro, J.C. (eds.) Euro-Par 2017. LNCS, vol. 10417, pp. 358–371. Springer, Cham (2017). https://doi.org/10.1007/978-3-319-64203-1_26
11. Lui, V.: Penetration testing: the white hat hacker. ISSA J. (2007)
12. McDermott, J.P.: Attack net penetration testing. In: Proceedings of the 2000 Workshop on New Security paradigms, pp. 15–21. ACM Digital Library (2001)
13. Williams, N., Zander, S., Armitage, G.: A preliminary performance comparison of five machine learning algorithms for practical IP traffic flow classification. ACM SIGCOMM Comput. Commun. Rev. **36**(5), 5–16 (2006)

14. Valea, O., Oprişa, C.: Towards pentesting automation using the metasploit framework. In: IEEE International Conference on Intelligent Computer Communication and Processing (ICCP), pp. 171–178 (2020)
15. Al-Shiha, R., Alghowinem, S.: Security metrics for ethical hacking. In: Arai, K., Kapoor, S., Bhatia, R. (eds.) SAI 2018. AISC, vol. 857, pp. 1154–1165. Springer, Cham (2019). https://doi.org/10.1007/978-3-030-01177-2_83
16. Verbedeke, A.: International Business Strategy: Rethinking the Foundations of Global Corporate Success. Cambridge University Press, Cambridge (2013)
17. Wang, J., Neil, M., Fenton, N.: A Bayesian network approach for cybersecurity risk assessment implementing and extending the FAIR model. Comput. Secur. **89**, 101659 (2020)
18. Raychaudhuri, S.: Introduction to Monte Carlo simulation. In: 2008 Winter Simulation Conference, Miami, FL, USA, pp. 91–100. IEEE (2008)

Archived Processes Integration for Extension of the Reverse Engineering Approach

Wafa Guefrech[1]([✉]), Sahbi Zahaf[1,2] [iD], and Faiez Gargouri[1]

[1] HICM, MIRACL Laboratory, University of Sfax, 242-3021 Sfax, Tunisia
faiez.gargouri@isims.usf.tn
[2] Higher Institute of Computer Sciences, University of Tunis El-Manar, Tunis, Tunisia
sahbi.zahaf@isi.utm.tn

Abstract. Enterprise Information System (EIS) must cover the interoperability criteria inside its "application view" scope. Nevertheless, the urbanization approach, on which we rely to implement this EIS, has to deal with "horizontal fit" and "traversal fit" problems: lack of intra and inter-applicative communications problems. To overcome these deficiencies, we show in this paper our solutions to ensure semantic interoperability between the applications of different EIS, which are involved to exploit collaboratively business processes. For that, we rely on the Semantic Web Services (SWS) to allow enterprises that use multiple software systems for a business management project, to exchange easily business data between these services. In this context, we propose to reuse and extend the reverse engineering approach that serves to describe the SWS. Four phases permit to describe our contribution: extraction of relevant terms, analysis of the filtered terms, construction of Ontology Design Patterns (ODP) and archived processes. We demonstrate that our strategy enhances the semantic interoperability criteria between the enterprises.

Keywords: Enterprise Information System · Semantic interoperability · Semantic web services · Reverse engineering · Archived processes · SESMA · SEDA

1 Introduction

In a market characterized by stiff competition, the enterprise ensures its survival through business that it achieves and bids that it wins. Thus, the strategic management of business is a major priority for an innovative enterprise, which promotes to restructure its information system around its trades and business processes. In fact, this management covers several processes (sales, purchasing, production, etc.). Each process exploits a specific software, which materializes its own graphical interface and its own database. In this case, the information disperses in disjoint systems ("spaghetti effect"), which reduces the productivity and the performance of each enterprise [24].

The EIS (Enterprise Information System) which allows exploiting Business Processes must be [22]:

© Springer Nature Switzerland AG 2021
I. Saad et al. (Eds.): ICIKS 2021, LNBIP 425, pp. 139–151, 2021.
https://doi.org/10.1007/978-3-030-85977-0_11

- Integrated: able to restore and exploit the patrimony of knowledge and expertise acquired during past experiments business.
- Flexible: able to resist to changes in the market and to cope with the agility of business.
- Interoperable: able to exploit communications between enterprises in order to contribute into the same project.

The urbanization approach, on which we rely to implement our EIS concerns four levels: Business View – representing the modeling of the business processes used by the enterprise; Functional View– representing the functions and information flows towards the business processes, regardless of the technologies used; Application View – representing the applications used to support functions and flows, and to facilitate the processes; Physical View – representing the "material" infrastructure.

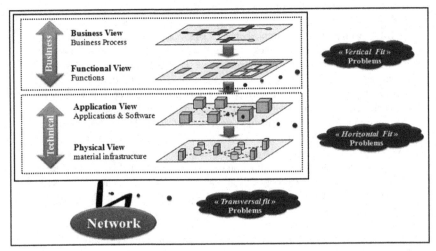

Fig. 1. Enterprise Architecture Information System reference model [7]: "three fit" problems [22].

The "vertical fit" represents the problems of integration from a business infrastructure, which is abstract, to a technical infrastructure, which represents implementations (lack of flexibility and lack of internal interoperability). The "horizontal fit" translates not only the software's problems of identification (induced by the "vertical fit" problems) that can cover the entire infrastructure of the enterprise's business, but also the intra-applicative communications problems (internal interoperability); the purpose being to ensure the interactions between software's of the same technical infrastructure in the enterprise (lack of flexibility and lack of internal interoperability). The "transversal fit" translates the inter-applicative communications problems: external interoperability carried out dynamically through a network (lack of openness and lack of external interoperability) [24].

We are particularly interested to propose our contributions to ensure the semantic interoperability criteria of any EIS inside its "application view" scope (Cf. Fig. 1). In this context, we propose to reuse and extend the reverse engineering approach [3] that serves to describe the SWS, which take as target a business process (WS–BPEL) [13] annotated by SESMA [11].

Four phases permit to describe our contribution. At the first and the second phases, we proceed to identify and analyze the filtered relevant terms that compose the business process ontology. At the third phase, we propose to build the Ontology Design Patterns (ODP) [22], which materialized the language of exploitation of the business process in a specific context. In the fourth phase we propose to integrate archived processes.

We demonstrate that our strategy enhances the semantic interoperability criteria between the enterprises.

This work structures as follows. The second section shows the related work that covers the semantic interoperability criteria. The third section describes our contribution in Enterprise Information System to deal with semantic interoperability problems. We end this work by the conclusion and with the prospects of our future works.

2 Related Work

The lack of interoperability appears in three levels in relation to the "three fit" problems. Thus, we distinguish:

- Business interoperability problem that is due to the heterogeneity of the business processes (design, language and formalism of representation).
- Technical interoperability problem that is due to the heterogeneity of the applications used for management of the business enterprise's features.
- Semantic interoperability problem that is due to the different types and formats of the data used by these applications. In fact, the semantic interoperability ensures if any application of a distributed heterogeneous system can exchange data and functionality easily with another one.

In the literature, there are different approaches that are interested in solving semantic interoperability problems. Especially, we are involved on the approach via ontology [16]and the approach via service [17]. In fact, the use of ontology allows the building a common standardized vocabulary within the organization and guarantees both formalization and sharing of the knowledge. This approach facilitates the communication between different actors involved in the business projects and leads to interoperability. Moreover, the use of service, particularly, the SOA (Service Oriented Architecture) [17] helps developing an easily flexible, extensible and adaptable EIS, which materializes by a set of reusable applications, blocks and refer the implementation of "services". These two approaches facilitate [21]: (1) the intra-applicative communications (ensure the interactions between applications of the same EIS), and (2) the inter-applicative communications (ensure the interactions between applications of different EIS). We can deduce that these approaches participate in the resolution of both "horizontal fit" and "transversal fit" problems.

Afterwards we are particularly interested to the technology Semantic Web Services (SWS) [16] that covers these two approaches. It is the result of the convergence between Web Services and the Semantic Web. Indeed, we rely on SWS to allow enterprises that use multiple software systems for business management to exchange easily business data over the internet. This technology facilitates the data transfer and communication between the services of different enterprises, which are involved in the same business project; thus it participates at solving "horizontal fit" and "transversal fit" problems [22].

Many approaches serve to describe SWS in order to improve their expressiveness, and consequently facilitate the process of automating their life cycle, such as: description languages, annotation languages, reverse engineering approach:

Description language: bases on the development of a complete semantic description language of web services and it derives from the semantic web. SWS like OWL-S (Semantic Markup for Web Services) [10, 15] and WSMO (Web Service Modeling Ontology) [12, 14] are part of this method.

Annotation language: bases on the existing languages annotations, benefiting of their extensibility advantages to improve and enrich their expressivity. These annotations may relate to:

- Business processes such as the SESMA (SEmantic Service Markup) approach [11], which permits to describe composite services by annotating WS-BPEL business processes [13].
- UDDI and ebXML directories using the SAWSDL (Semantic Annotation for WSDL) [12, 19].
- DAML-S (DARPA agent markup language for services) [18].
- WSDL (Web Service Description Language) [20] documents such as the DAML-S languages, the USDL (Handbook of Service Description) [9].
- Reverse engineering [4]: permits to identify for any system studied (i.e., business process), its concepts and their correlations. This approach relies on conceptual diagram, which is rich in semantic area and permits to describe the internal functioning or manufacturing method for this system (i.e., business process). Thus, we can rely to this approach to describe web services in order to facilitate its: maintenance, evolution and understanding [3, 4].

Table 1 shows the comparative study between these three methods that covers the SWS technology according the semantic interoperability area. Different criteria are proposed by [5] to do this comparison: dependence throws ontology language, semantic explicitness, adaptability and reuse, and expressivity. We propose to extend this comparison study by added the "business coverage" criteria, which evaluates the capacity of a description language to annotate a business process, and thus to cover all business needs of the EIS.

For this comparative study (Table 1), we can deduce there is no approach, which can cover all these criteria's. Each approach has its advantages and its limits. In order to reap the benefits of these approaches and, thus to cover their gaps, we propose to core of the coupling capacity of the reverse engineering approach by using SESMA language [11]. This premise permits to enhance and provide a rich quality description of SWS.

The majority of reverse engineering approaches are extensible, adaptable and support the rich semantics models. They based on the ontological solutions, which provide to cover any business domain by specifying a set of terms and their relationship's.

However, some of these solutions are not generic and depend on ontology languages; even they don't reflect the reality at the exploitation level, insofar as, they are different from the targets. Some ontological solutions take as a target a conceptual graph [1]; others a HTML and XML documents [21], or the source code of web service; and the rest a specific service description language (like the researches proposed on [2] and [3]). Note that, the extensibility of this last solution permits to improve the expressivity (like SESMA [11] for example). Indeed, SESMA has the advantages related to the other languages of annotation and description of web services; it permits the annotation of business processes, which is a strong point to reflect the business needs constituting the service.

Table 1. Comparative study between the SWS methods according the semantic interoperability area.

	Approaches	Description approaches based on semantic languages		Description approach based on annotation			Reverse engineering approach(RE)	
	Criteria	OWL-S	WSMO	SAWS DL	USDL	SESMA	Description and classification of SWS WSMO applications	Web Application Based on Domain Ontology
Semantic Interoperability	Dependence throws ontology languages	-	-	++	-	+	-	+/-
	Semantic explicity	+	+	-	+/-	+/-	+	+
	Expressivity	+/-	+/-	+	+/-	+	+/-	+/-
	Adaptability and Reuse	+	+	++	-	++	++	++
	Business coverage	+/-	+/-	+/-	+/-	++	+/-	+/-

Our goal consists to support a semantic description for web services in the perspective to ensure semantic interoperability. For that, we propose to extend the reverse engineering approach by an extraction process of relevant terms. Furthermore, we propose to support our contribution by adding an annotation language to increase its expressiveness and to improve its semantics. This proposition permits to reflect the concrete business needs of each organization and to facilitate user's tasks. The next section shows the details of our contribution.

Enterprise Information System (EIS) must cover the semantic interoperability criteria inside its "application view" scope, which represents the intra-applicative communications between software's of the same EIS used to support the business processes (Cf. Fig. 1).

3 Contributions Addressing the Semantic Interoperability Criteria for the EIS

The works of [23] address integrity, interoperability and flexibility by defining four dimensions to describe the EIS: OPERATIONAL DIMENSION – serves to exploit business processes used by the enterprise in its different projects; ORGANIZATIONAL DIMENSION – organizes the expertise that the enterprise acquired during its previous projects. Such assumption permits an eventual adaptation and reutilization of these skills in future projects; DECISION-MAKING DIMENSION – aims to make the right decisions that concern the enterprise participation in different projects. COOPERATIVE DIMENSION – covers the intra-enterprise communications and the inter-enterprise communications in order to contribute on the techno-economic bid proposition. Table 2 illustrates the different approaches used to describe these four dimensions EIS.

Table 2. EIS in the core of the coupling capacity of the eight approaches [23]

		Flexibility		
		Integrity		External Interoperability
		Internal Interoperability		
		"Vertical fit"	"Horizontal fit"	"Transversal fit"
Operational Dimension	Organizational Dimension	PRIMA / Engineering Systems / Lean Manufacturing / SOA /BPM	Knowledge Management	Knowledge Management
	Decision Dimension		Business Intelligence	
	Cooperative Dimension		SOA / Web Service	SOA / Web Service / Cloud Computing

Table 2 demonstrated that Knowledge Management (KM), Service Oriented Architecture (SOA) and Web Services (WS) allow us to overcome "horizontal fit" and "transversal fit" problems [22]:

- KM allows the organization to formalize its business knowledge during different business projects that it realizes. This approach facilitates establishing a pivot language between companies and participates in solving the problem of "horizontal fit" and "transversal fit".
- SOA allows the organization to develop easily flexible, extensible and adaptable EIS. This approach standardizes the intra-applicative communications (SOA participates in the resolution of the problem of "horizontal fit"), as well as inter-applicative (SOA participates in solving the problem of "transversal fit").

- WS (Web Service) allows organization that uses different software's systems for business management to exchange business data over the internet. This approach facilitates the data transfer between enterprises and participates in solving the problem of "horizontal fit" and "transversal fit".

The use of these approaches permits to guarantee the semantic interoperability criteria of the enterprise. It allows solving the problems of heterogeneity that inhibit the establishment of communications, both inside the scope of an EIS and between different EIS.

We have demonstrated in the previous section, that Semantic Web Services (SWS) [17] covers these approaches. That is why we propose to support "application view" of an EIS with this technology, which allows different applications to operate and communicate from the web. However, these SWS must described in order to facilitate and automate actions related to their life cycle and allow machines to interpret them. For example, the reverse engineering approach serves to describe SWS through the documentation; it allows a better exploitation of the EIS. However, this approach has certain failures such as the incompleteness of the specification of relevant terms and the dependence on ontology description language. In order to deal with this luck of expressivity, we propose to present our generic solution that is not depend on web services languages, but it is based to reuse and adapt the business needs by using other technologies to bring more semantics. Concretely, our contribution is an extension of the reverse engineering approach; it described according to three phases (Cf. Fig. 2).

- Extraction of relevant terms: the extension that we propose consists to apply the filtering technique, like the extraction of relevant knowledge process proposed by [22]. Particularly, we propose to use as target a business process (WS–BPEL [13]) annotated using SESMA language (see Fig. 2). Note that, SESMA allows facilitating the tags that it defines for the description, through its syntax.
- Analysis of the extracted terms: the filtered terms compared with the concepts of domain ontology. We use the same metrics proposed on [3] and [4].
- Construction of Ontology Design Patterns (ODP): in this new phase, we propose to generate a language that materialized the exploitation of the business process in a specific context. This solution permits to solve problems related to contextual conflicts [22]. Indeed, the use of an ODP facilities the reutilization of the business process.

3.1 Extraction of the Relevant Terms Supported by the Filtering Technique

In order to enhance the expressivity of the Reverse Engineering, we propose to extend it by enriching the process of extraction of relevant terms (Cf. Fig. 3). This process consists to determinate the relevant terms from a document.

In our case, we used a SESMA document, which allows in addition, the description of services and the annotation of business processes WS-BPEL that translating the business needs. In fact, in order to have a rich description of services; we opt to do a simulation of this process by developing a web application that extract the relevant terms based in SESMA document [8]. The SESMA document will undergo the following steps:

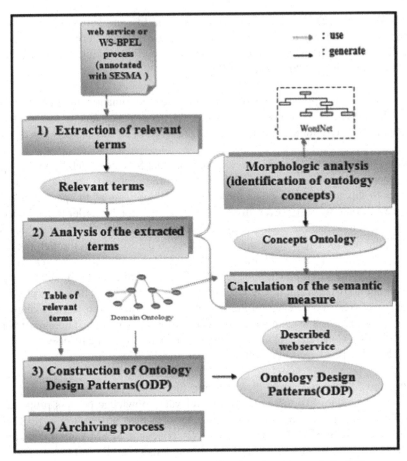

Fig. 2. Our extension of the Reverse Engineering Approach [3, 4, 8].

- Cleaning Process: permits to verify the consistence of the document. The result of this step is a set of cleaned document.
- Filtering and analysis process: consists to brows its source code, eliminate unnecessary tags such as formatting one and permits to keep only the important and the useful tags such as: <input> <output>, <precondition>, <effect>, etc. and keep only the tags. The result of this step is a set of filtered and analyzed documents.
- Inference of relevant terms: permits to extract from filtered and analyzed document the relevant terms.
- Capitalization of relevant terms: consists to save and store extracted terms on tables, in order to facilitate their reuse, it in the process of construction of Ontology Design Pattern.

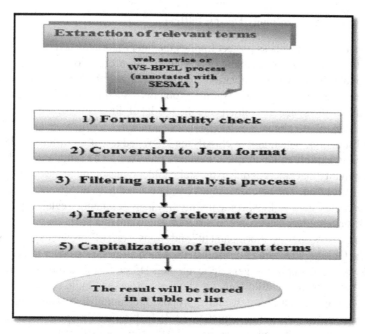

Fig. 3. Our process extraction of the relevant terms [8].

3.2 Analysis of the Filtered Terms

In this phase, we use the same steps followed by [3, 4]. We propose to analyze the relevant terms, which extracted and filtered in the previous phase. Figure 2 shows that this phase is composed of two steps:

- Morphologic analysis (identification of ontology concepts): consists to extract the concepts of domain ontology (that are related to each other and cover a specific domain). Our objective consists to compare these concepts with the relevant terms extracted in the previous phase, in order to know if the web service and the ontology are parts of the same domain and allow us to take this ontology as reference to describe the web service.
- Calculation of the semantic measure: consists to calculate the similarity between the concepts of domain ontology and the relevant terms.

This methodology used to calculate the semantic distance of ontological reverse engineering: (1) we have a chart corresponding to a web service; (2) we have ontology of domain containing concepts: each concept can possibly have a set of attributes; (3) we calculate semantic distance to can identify the SWS domain and can import a new Ontology Design Patterns.

3.3   Construction of Ontology Design Patterns

To work on a specific business project implies the intervention of several collaborators. Certainly, these contributors exchange knowledge and information flows. However, its environmental differences lead to various representations and interpretations of knowledge: ("horizontal fit" and "transversal fit" problems). Such failures described in terms of five conflicts: the syntactic conflicts are the results of different terminologies used by stakeholders on a same domain of business. The structural conflicts related to different levels of abstraction, which aim at classifying know-ledge within different EIS, which are involved to exploit collaboratively business processes. The semantic conflicts concern the ambiguity that emerges due to the stakeholders' reasoning in the development of the business proposal. The heterogeneities conflicts are due to the diverse data sources. Finally, the contextual conflicts, come mainly from environmental scalability problems, and in fact, stakeholders can evolve in different environments [23].

In order to overcome the "horizontal fit" and "transversal fit" problems, we propose to generate the Ontology Design Patterns (ODP), which allows us to cover the complexity of consensual modeling at the generic level. This solution solves problems related to contextual conflicts. In the following, we show our process to build an ODP.

At the end of the first phase, we have proposed to capitalize and store the relevant terms in tables, in order to exploit them (Cf. Fig. 3). In this phase, we propose to reuse only the terms that have a high level of similarity with the concepts of domain ontology. These terms allow us to build the Ontology Design Patterns (ODP) [22], which materialized the language of exploitation of the business process in a specific context. Figure 4 shows our method of construction of the ODP.

This solution permits to share and reuse knowledge's between actors, which are involved in the collaborative exploitation of the business process. In addition, it allows us to refine and improve the domain ontology.

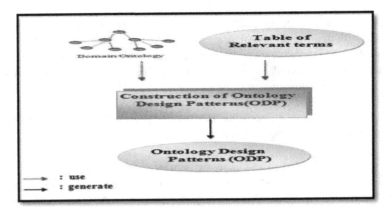

Fig. 4. Our method of Construction of the Ontology Design Patterns (ODP) [8].

3.4 Archived Processes

With the increase in the amount of stored data, the company must confront this saturation by deploying a strategic method of organization which takes into account the development of the electronic administration in order to facilitate the reuse in future projects and the share with company stakeholders since they contain knowledge and know-how [6].

By the way, the development of electronic administration plays an important facilitating role in improving the performance of the information system by enhancing the exchange of data and the automation of administrative processes.

Here appears the importance of the archived processes that aims to preserve information to allow users to share and use it.

In fact, the implementation of the archived processes requires the use of a generic and reusable model to facilitate interoperability between the archiving service information system and the information systems of its partners in the context of their exchanges of data.

In this way, SEDA (Data Exchange Standard for Archiving) [6] aims to facilitate exchanges between information systems by guaranteeing interoperability.

Also, it constitutes a strong structuring element for the services that they use it insofar as the formalism used makes it possible to finely control the exchanges.

This standard is registered in the general interoperability repository where it is included and cited as recommended.

Also, SEDA is based on already existing norms and standards such as; ISO 14 721 (OAIS model) which constitutes its basic structure and the xml language which was chosen to structure the information contained in the exchange standard.

Given the importance of this process we proposed to integrate it into our approach in order to further support the interoperability of EIS.

4 Conclusion and Perspectives

In this paper, we have proposed our contribution to ensure the semantic interoperability criteria between the applications of the EIS of different enterprises, which are involved to exploit collaboratively business processes. For that, we have relied on the SWS to exchange easily business data between these enterprises. Concretely, we have proposed to reuse and extend the reverse engineering approach, which takes as a target a SESMA document. We described our contribution according three phases:

First, we proceeded at identifying the filtered relevant terms and we proposed to simulate it by implementing an extraction tool of relevant terms.

Second, we analyzed these terms with the concepts of the business ontology domain; Third, we proposed to build an ODP with terms that present a high degree of similarity compared to the concepts domain. Our approach permits to generate different ODP, which are involved at the exploitation of the business process in a specific context. Thus, we have proposed solutions to deal with "horizontal fit" and "transversal fit" problems, on which we rely to implement an EIS. Also we proposed to integrate the archived processes in our approach that aims to facilitate exchanges between information systems by guaranteeing interoperability.

In future work we suggest exploiting our contribution to generate the ODP that permit to exploit the bid process. It aims to examine the feasibility of the bid before negotiating any contract with any owner following a pre-study carried out before a project launch. This process translates the techno-economic expertise; which partners build in a cooperative way. It is a key business process, which influences the company's survival and strategic orientations [22]. Also we suggest to extend the standard of exchange in order to enhance its function.

References

1. Antoniol, G., Canfora, G., Casazza, G., De Lucia A.: Web site reengineering using RMM. In: 2nd International Workshop on Web Site Evolution, 1 March 2000
2. Bell, D., de Cesare, S., Lycett, M.: Semantic Transformation of web services. In: Meersman, R., Tari, Z., Herrero, P. (eds.) OTM 2005. LNCS, vol. 3762, pp. 856–865. Springer, Heidelberg (2005). https://doi.org/10.1007/11575863_107
3. Bedad, F., Haouas, A., Bouchiha, D.: Description et Classification des Services Web Sémantiques, Nature & Technologie (2012)
4. Bouchiha, D., Mimoun, M., Mohamed, B.S., Maamar, K.: La rétro- ingenierie des applications Web : une approche basée sur l'ontologie de domaine (2006)
5. Chabeb, Y.: Contribution à la description et la découverte des Services Web Sémantiques. Thèse de doctorat en Informatique de Télécom SudParis dans le cade de l'école doctorale S & I en co-accréditation avec l'Université d'Evry-val d'Essonne (2011)
6. Sibille-de Grimoüard, C., Nichèle, B.: «Le Standard d'échange de données pour l'archivage (SEDA), un outil structurant pour l'archivage». In : La Gazette des archives, no. 240, pp. 153–164 (2015)
7. Fournier-Morel, X., Grojean, P., Plouin, G., Rognon, C.: SOA le Guide de l'Architecture du SI. Dunod, Paris (2008)
8. Wafa, G., Zahaf, S., Gargouri, F.: Semantic interoperability for enterprise information system. In: 11th Conference on ENTERprise Information Systems, Sousse, Tunisia, 16–18 October 2019. Elsevier Ltd. (2019)
9. Handbook of Service Description: USDL and its methods
10. Izza, S., Vincent, L., Burlat, P.: Exploiting semantic web services in achieving flexible application integration in the microelectronics field. Comput. Ind. **59**(7), 722–740 (2008)
11. Peer, J.: Semantic service markup with SESMA, MCM Institute University of St. Gallen Blumenbergplatz 9 9000 St. Gallen
12. Mecheri, K., Boufaida, M., Souici, H., Meslati, D.: Etude des mécanismes d'interopérabilité des systèmes d'information basés sur les services Web sémantiques, Rev. Sci. Technol. Synthèse (2017)
13. OASIS standard, OASIS: Web Services Business Process Execution Language (WS- BPEL) version 2.0 (2006)
14. Roman, D., et al.: Web Service modeling ontology. Appl. Ontol. J. **1**(1), 77–106 (2005)
15. Sabou, M., Wroe, C., Goble, C., Stuckenschmidt, H.: Learning domain ontologies for semantic web service descriptions. J. Web Semant. **3**, 340–365 (2005)
16. Gruber, T.R.: Towards principles for the design of ontologies used for knowledge sharing in formal ontology in conceptual analysis and knowledge representation. Kluwer Academic Publishers (1993)
17. Wu, Z., Palmer, M.: Verb semantics and lexical selection. In: 32nd Annual Meeting of the Association for Computational Linguistics, pp. 133–138 (1994)

18. W3C standard DAML-S: Semantic Markup for Web Services. http://www.daml.org/services/daml-s/0.9/daml-s.html
19. W3C standard, Semantic Annotations for WSDL and XML Schema (SAWSDL) is a 2007 published technical recommendation of W3C DARPA agent markup language for services (DAML-S)
20. W3C standard, WSDL ou Web Services Description Language (2001). La version 2.0 a été approuvée le 27 juin 2007
21. Yip Chung, C., Gertz, M., Sundaresan, N.: Reverse engineering for web data: from visual to semantic structures. In: Proceedings of the 18th International Conference on Data Engineering (ICDE'02) (2002)
22. Zahaf, S., Gargouri, F.: ERP Inter-enterprises for the operational dimension of the urbanized bid process information system. In: 6th Conference on ENTERprise Information Systems, vol. 16, pp. 813–823 (2014). J. Procedia Technol. Elsevier Ltd, 15–17 Octobre 2014, Troia, Portugal
23. Zahaf, S., Faiez, G.: The urbanized bid process information system. In: 21th International Conference in Knowledge Based and Intelligent Information and Engineering Systems, vol. 112, pp. 874–885 (2017). Journal of Procedia Computer Science, Elsevier Ltd, 06–08 Septembre, 2017, Marseille, France
24. Zahaf, S., Gargouri, F.: Specification for the cooperative dimension of the bid process information system. In: 9th Conference on ENTERprise Information Systems, vol. 121, pp. 1023–1033 (2017). Journal of Procedia Computer Science, Elsevier Ltd, 08–10 Novembre, 2017, Barcelone, Spain

QoS Aware Classification of Composite Web Services Using Rough Approximation

Salem Chakhar[1,2(✉)], Ahmed Abubahia[3], and Farok Bin Iqdara[1]

[1] Portsmouth Business School, University of Portsmouth, Portsmouth, UK
salem.chakhar@port.ac.uk, Farok.BinIqdara@myport.ac.uk
[2] Centre for Operational Research and Logistics, University of Portsmouth, Portsmouth, UK
[3] School of Psychology and Computer Science, University of Central Lancashire, Lancashire, UK
AAbubahia@uclan.ac.uk

Abstract. A typical Web service composition approach contains four phases: (1) construction of the compositions, (2) generation of executable plans, (3) classification of executable plans, and (4) selection and deployment of the best executable plan. This paper adopts and extends a graph based composition approach by using a more reliable classification method, namely Dominance-based Rough Set Approach (DRSA). This method is very suitable to take into account the Quality of Service (QoS) aspects in Web service composition. The paper also illustrates the application of the DRSA on a set of composite Web services and evaluates its performance.

Keywords: Web service composition · Quality of Service · Dominance-based rough set approach · Classification · Decision making · Artificial intelligence

1 Introduction

Web services are self-contained and self-describing application components that can discovered and invoked by other applications. The basic Web service architecture contains three elements: (i) *service requester*, which is the software system that requests, (ii) *service provider*, which is the software system that would process the request and provides the data, (iii) *service registry*, which contains additional information about the service provider. Composing individual Web services to construct new and more complex Web services is a current solution to deal with complex situations in organisations. The input to Web service composition is a set of specifications describing the capabilities of the desired service. These specifications are decomposed into two groups: (i) functional requirements that deal with the desired functionality of the composite service, and (ii) non-functional requirements that relate to issues like cost, performance and availability. These specifications need to be expressed in an appropriate language.

© Springer Nature Switzerland AG 2021
I. Saad et al. (Eds.): ICIKS 2021, LNBIP 425, pp. 152–167, 2021.
https://doi.org/10.1007/978-3-030-85977-0_12

For instance, functional requirements can be expressed using the Ontology Web Language (OWL) and non-functional requirements in terms of Quality of Service (QoS).

Different Web service composition approaches have been proposed in the literature [9,12,15–17]. An important issue within Web service composition is the selection of the most appropriate one among the different possible compositions using both functional and non-functional requirements. The functional requirements of a Web service are generally expressed through Inputs-Outputs-Preconditions-Effects (IOPE), which captures the transformation performed by this service. The non-functional requirements represent desirable capabilities that cannot be correctly handled through IOPE. A possible solution is to use the QoS to evaluate, compare and select the best composition.

The end-to-end composition operation starts by user specification of functional and non-functional requirements and leads to an executable plan that can be handed off to runtime environment for execution. In this paper, we adopt and extend a graph based composition approach proposed in [1,3]. This approach is organised into four phases: (1) construction of the composition graph, (2) generation of executable plans, (3) classification of executable plans, and (4) selection and deployment of the best executable plan. In [3], the classification phase relies on ELECTRE TRI method that is very demanding in terms of preference and technical parameters. In this paper, we propose to use a relatively recent and more appropriate classification method, namely the Dominance-based Rough Set Approach (DRSA) [7,14].

The rest of the paper is organised as follows. Section 2 reviews related work. Section 3 introduces compositions construction strategy. Section 4 applies DRSA to evaluate and classify compositions. Section 5 presents some implementation and evaluation issues. Section 6 concludes the paper.

2   Related Work

The authors in [16] categorise Web service composition into three core research concerns: service classification, planning and selection. In this section, we mainly focus on the classification of composite Web services, which is a crucial step in all composition approaches. Early composition approaches use purely functional requirements to select among composite Web services, while more recent composition approaches consider both functional and non-functional requirements, permitting thus to handle QoS aspects [12,16,19].

Most of existing QoS aware proposals rely on the use of a successive evaluation of non-functional aspects in order to attribute a general level of quality to different composite Web services and to select the 'best' one from these services. In these works, the evaluation of composite Web services is based either on a single attribute or on a weighted-sum of several quantitative attributes. Other techniques include linear programming, which is used to compute an 'optimal' execution plan for composite Web service. More recent approaches rely on deep neural network [17] and MapReduce based evolutionary algorithm [10]. All these techniques cannot handle correctly the QoS aspects.

To overcome the shortcomings of the above cited approaches, several authors proposed the use of multicriteria techniques [1,3,4]. For instance, the author in [1] proposed a graph based and QoS enhanced Broker for Web service composition that uses ELECTRE TRI method to classify composite Web services. Other composition approaches, including [11,18], rely on the use of classical Indiscernibility-based Rough Set Approach (IRST) [13]. For example, in [11], the authors integrated IRSA in cloud service selection mechanism in order to provide right decisions for cloud users and efficient service improvement for cloud providers. The IRSA is well adapted to deal with functional requirements, but fails to handle the preference-related aspects of QoS attributes.

In this paper, we support the use of the DRSA. This method extends classical IRSA to multicriteria classification. The DRSA has several attractive characteristics as it: (i) does not need any preference parameters, which reduces the cognitive effort required from the service client; (ii) produces if-then decision rules, which are easily understood by the user; (iii) is able to deal with incomplete/missing attribute values; and (iv) is able to detect and deal with inconsistency problems.

3   Construction of Composite Web Services

In this section, we briefly review the construction of the composite Web services in [3]'s composition approach. The extended classification phase will be detailed in Sect. 4. The selection and deployment of the best executable plan is beyond the scope of this paper.

3.1   Service Type and Instance

This paper distinguishes between Web service *types*, which are groupings of similar (in terms of functionality) Web services, and the actual Web service *instances* that can be invoked. The Web service types and instances can be advertised in a registry. A **Web service type** S_i is a tuple $\langle F_i, Q_i, H_i \rangle$, where:

- F_i is a description of the service's functionality,
- Q_i is a specification of its QoS attributes, and
- H_i is its cost specification.

Although the cost specification is often considered as a QoS attribute, it is here treated separately since it usually represents a major decision attribute for service selection. Each Web service type S_i has a unique functionality F_i. In turn, the same functionality may be supported by different service types. Let $I_i = \{I_1^i, I_2^i, \cdots, I_{n_i}^i\}$ denotes the set of service instances associated with service type S_i where n_i is the number of instances for service type S_i.

3.2   Composite Service Type and Instance

A **composite service type** is a tuple $\langle S, R, Q, H \rangle$, where:

- $S = \{S_1, \cdots, S_n\}$ is a collection of n service types,
- R is a specification of the invocation relationships in S,
- Q is a specification of its QoS attributes, and
- H is its cost specification.

The functionality of the composite service can be retrieved from set R. Naturally, the specification Q of QoS attributes and cost specification H of a composite service are defined based on the QoS attributes Q_i and cost H_i specifications of simple services considered in the composition. Accordingly, appropriate aggregation rules need be defined and used to combine QoS and cost attributes of simple services into specifications that apply to the composite service as a whole. This issue is addressed in Sect. 3.4.

Let $J = \prod_{\alpha=1}^{n} I_\alpha$ be the set of composite service instances. Let us assume that $J = \{J_1, \cdots, J_k, \cdots, J_m)$. A **composite service instance** $J_k \in J$ ($k = 1, \cdots, m$) is a collection of service instances $(I_.^1, \cdots, I_.^i, \cdots, I_.^n)$ with $I^i \in I_i$ for $i = 1$ to n.

We note that the same composite service type may be implemented by different composite service instances.

3.3   Composition Graph

A composition operation involves several individual Web services. The relationships among the individual Web service types in S can be represented by a connected and directed graph $G = (X, V)$ where $X = \{S_i, S_j, \cdots, S_m\}$ is the set of individual Web services and $V = \{(S_i, S_j) : S_i, S_j \in X \wedge S_i \text{ can invoke } S_j\}$. $G = (X, V)$ is called the **composition graph**. Figure 1 presents a composition graph example implying six individual Web services S_1, S_2, S_3, S_4, S_5 and S_6.

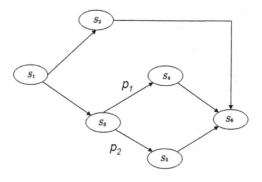

Fig. 1. Example of composition graph.

The arcs in this figure represent different types of invocation that can de modelled through Business Process Execution Language (BPEL) constructors. The basic BPEL constructors are: sequential, parallel, probabilistic, conditional and synchronised. In *sequential invocation*, a Web service is activated as a result of the completion of one of a set of mutually exclusive predecessor activities. The *parallel invocation* (fork) represents a point in the process where a single thread of control splits into multiple threads of control which can be executed in parallel. In the *probabilistic invocation*, a probability value p on an outgoing arrow from S_i to S_j indicates that S_i invokes S_j with probability p. For example, Web services S_4 and S_5 in Fig. 1 are invoked with probabilities of p_1 and p_2, respectively. The *conditional invocation* represents a situation where one or several branches are chosen. In the *synchronised invocation* (join), a Web service is activated only when all of its predecessor Web services have completed.

Let P_i be the collection of providers supporting functionality F_i of Web service S_i: $P_i = \{s_i^1, s_i^2, \cdots, s_i^{n_i}\}$ where n_i is the number of providers in P_i. A composite Web service is then defined as follows. Let $S_1, S_2, \cdots, S_n$ be a set of n individual Web services such that $S_i = (F_i, Q_i, H_i)$ ($i = 1, \cdots, n$). Let $G = (X, V)$ be the composition graph associated with $S_1, S_2, \cdots, S_n$. A **composite Web service** k is an instance $\{s_1, s_2, \cdots, s_n\}$ of G defined such that $s_1 \in P_1$, $s_2 \in P_2$, $\cdots$, $s_n \in P_n$.

To take into account the invocation probabilities associated with some Web services, a new function, called π, is defined as follows:

$$\pi: X \times X \to [0, 1]$$
$$S_i \times S_j \to \pi(S_i, S_j)$$

The number $\pi(S_i, S_j)$ represents the probability that S_i invokes S_j. For illustration, let us consider the graph of Fig. 1 and suppose that: $P_1 = \{s_1^1, s_1^2, s_1^3, s_1^4\}$, $P_2 = \{s_2^1, s_2^2, s_2^3\}$, $P_3 = \{s_3^1, s_3^2\}$, $P_4 = \{s_4^1\}$, $P_5 = \{s_5^1, s_5^2, s_5^3, s_5^4, s_5^5\}$ and $P_6 = \{s_6^1, s_6^2\}$, where s_i^j is the jth provider of Web service S_i. According to the definition of a composite Web service given earlier, we can construct 240 composite Web services from the composition graph in Fig. 1. For instance, composite Web service k_{49}, associated with $G_{49} = (X_{49}, V_{49})$, is specified as follows:

- $X_{49} = \{s_1^1, s_2^3, s_3^1, s_4^1, s_5^5, s_6^1\}$.
- $V_{49} = \{(s_1^1, s_2^3); (s_1^1, s_3^3); (s_2^3, s_6^1); (s_3^1, s_4^1); (s_3^1, s_5^5); (s_4^1, s_6^1); (s_5^5, s_6^1)\}$.
- $\pi(s_3^1, s_4^1) = p_1$; $\pi(s_3^1, s_5^5) = p_2$.

3.4   Partial Evaluation of Compositions

As defined earlier, a composition is an instance of the composition graph G. Each composition can be seen as collection of individual Web services. The evaluation provided by the UDDI registry are relative to these individual Web services. Thus, to compute the partial evaluations $q_j(k_i)$ ($j = 1, \cdots, m$) of the different compositions k_i ($i = 1, \cdots, n$), a set of m aggregation operators $\Phi_1, \Phi_2, \cdots, \Phi_m$, one for each QoS attribute, are required. The partial evaluation of a composition k_i on QoS attribute q_j, $q_j(k_i)$, is obtained by applying a bottom-top scan on graph $G_i = (X_i, V_i)$ and by applying the aggregation operator Φ_j on each node.

The valuation with respect to QoS attribute q_j of a node $x \in X_i$, denoted $v_j(x)$, is computed as follows:

$$v_j(x) = \Phi_j[q_j(x), \Omega(\Gamma^+(x))]$$

where $\Gamma^+(x)$ is the set of successors of node x. The operator Ω involves nodes on the same level and may be any aggregation operator such as sum, product, max, min, average, etc. The operator Φ_j concerns nodes on different levels and vary according to the BPEL constructors associated with node x. It may be the sum, product, max, min, or average.

It is important to note that when the QoS attribute is ordinal, it is not possible to use the probability associated with the branches of a <switch> constructor. To avoid this problem, one of the following rules may be used (other rules may also apply): (i) ignore the probabilities and proceed as with the <flow> BPEL constructor; (ii) use the partial evaluation associated with the most probable branch; (iii) use the majority rule (when there are at least three branches); (iv) use the intermediate level between the partial evaluations associated with most probable branch and least probable branch; and (v) use the intermediate level between the highest partial evaluation and the lowest partial evaluation.

An example illustrating the calculation of partial evaluations is given in Appendix A.

4   Classification of Composite Web Services

In this section, we apply DRSA to assign a set of compositions into a set of predefined and preference ordered QoS classes. The DRSA can be organised into three main steps: (1) specification of learning set, (2) approximation, and (3) induction, validation and exploitation of decision rules. More details about this approach are given in [7,14].

4.1   Learning Set Specification

The DRSA is a knowledge discovery method that takes as input a decision table $S = \langle U, C, D \rangle$ where U is the learning set, C a set of condition attributes and D a set of decision attributes defined such that $C \cap D = \emptyset$. It is assumed that the domains of attributes are preference-ordered and that the set $D = \{d\}$ is a singleton. The unique decision attribute makes a partition of U into a finite number of preference-ordered decision classes $Cl = \{Cl_t, t = 1, \ldots, h\}$, such that each $x \in U$ belongs to one and only one class.

For illustration, a collection of composite Web services defined based on Fig. 1 have been identified and used as learning set. Let assume that the specifications of QoS attributes used are those in Table 1. As shown in this table, QoS attributes q_1 and q_3 are to be minimised while QoS attributes q_2 and q_4 are to be maximised.

Table 1 also indicates that the three first condition attributes are cardinal. The latter is ordinal for which the following five-level scale ranging from 1 (very low) to 5 (very high) is used. The evaluations of the learning examples with respect to the QoS and decision attributes are shown in Table 2. The last column of this table indicates the values of decision attribute d, which partitions composite Web services into four QoS classes: $Cl_1 = \{d = 1\}$, $Cl_2 = \{d = 2\}$, $Cl_3 = \{d = 3\}$, and $Cl_4 = \{d = 4\}$ where $Cl_4 = \{d = 4\}$ is the most preferred QoS class.

Table 1. Characteristics of QoS attributes.

Code	Name	Description	Preference	Data type
q_1	Response Time	Time from request sending to response reception	Cost	Continuous
q_2	Availability	Degree to which the service is operable	Gain	Continuous
q_3	Cost	Web service cost specification	Cost	Continuous
q_4	Security	Level of security service provider	Gain	Ordinal

Table 2. Learning set.

#	Description	Response Time (q_1)	Availability (q_2)	Cost (q_3)	Security (q_4)	QoS Class
1	$\{s_1^1, s_2^1, s_3^2, s_4^1, s_5^1, s_6^1\}$	9.22	0.45946	2.48	1	3
2	$\{s_1^1, s_2^1, s_3^2, s_4^1, s_5^3, s_6^2\}$	8.12	0.66199	3.2	2	4
3	$\{s_1^1, s_2^2, s_3^2, s_4^1, s_5^3, s_6^2\}$	8.12	0.59512	3.1	2	4
4	$\{s_1^1, s_2^3, s_3^1, s_4^1, s_5^2, s_6^2\}$	8.12	0.41817	2.68	1	2
5	$\{s_1^2, s_2^2, s_3^2, s_4^1, s_5^1, s_6^3\}$	7.82	0.57107	3.2	2	4
6	$\{s_1^3, s_2^1, s_3^1, s_4^1, s_5^5, s_6^1\}$	9.12	0.38604	2.74	3	3
7	$\{s_1^3, s_2^1, s_3^2, s_4^1, s_5^2, s_6^2\}$	7.42	0.40317	3.38	2	2
8	$\{s_1^4, s_2^1, s_3^1, s_4^1, s_5^1, s_6^1\}$	10.92	0.44145	2.08	1	1
9	$\{s_1^4, s_2^2, s_3^2, s_4^1, s_5^3, s_6^2\}$	9.52	0.56506	2.8	2	2
10	$\{s_1^4, s_2^2, s_3^2, s_4^1, s_5^5, s_6^2\}$	10.12	0.53294	2.68	2	2
11	$\{s_1^4, s_2^3, s_3^1, s_4^1, s_5^1, s_6^2\}$	10.42	0.48356	2.32	1	1

4.2 Approximation

The DRSA relies on the dominance relation to approximate the decision classes in Cl. The *dominance relation* Δ is defined for each pair of objects x and y as follows:

$$x \Delta y \Leftrightarrow f(x, q) \succeq f(y, q), \forall q \in C. \qquad (1)$$

where $f(x,q)$ and $f(y,q)$ are the evaluations of x and y with respect to condition attribute q, respectively. Two sets are associated to each object $x \in U$: (i) the *dominating set* $\Delta^+(x) = \{y \in U : y\Delta x\}$ containing the objects that dominate x, and (ii) the *dominated set* $\Delta^-(x) = \{y \in U : x\Delta y\}$ containing the objects dominated by x.

Due to the preference order over Cl, the sets to be approximated are not single classes but collections of upward unions $Cl_t^{\geq} = \cup_{s \geq t} Cl_s$ and downward unions $Cl_t^{\leq} = \cup_{s \leq t} Cl_s$ of classes. The union $Cl_t^{\geq}$ is the set of objects that belong to at least class Cl_t, while the union $Cl_t^{\leq}$ is the set of objects that belong to at most Cl_t. Each class union $Cl_t^{\geq}$ (resp. $Cl_t^{\leq}$) will then be defined through its (i) lower approximation and (ii) upper approximation. The lower and upper approximations of $Cl_t^{\geq}$ are defined as follows:

- $\underline{P}(Cl_t^{\geq}) = \{x \in U : \Delta_P^+(x) \subseteq Cl_t^{\geq}\}$,
- $\bar{C}(Cl_t^{\geq}) = \{x \in U : \Delta^-(x) \cap Cl_t^{\geq} \neq \emptyset\}$.

The lower and upper approximations of $Cl_t^{\leq}$ are defined as follows:

- $\underline{C}(Cl_t^{\leq}) = \{x \in U : \Delta^-(x) \subseteq Cl_t^{\leq}\}$,
- $\bar{C}(Cl_t^{\leq}) = \{x \in U : \Delta^+(x) \cap Cl_t^{\leq} \neq \emptyset\}$.

The lower approximations group the objects which certainly belong to class unions $Cl_t^{\geq}$ (resp. $Cl_t^{\leq}$). The upper approximations group the objects which could belong to $Cl_t^{\geq}$ (resp. $Cl_t^{\leq}$).

The boundaries of $Cl_t^{\geq}$ and $Cl_t^{\leq}$ are defined as follows:

- $Bn(Cl_t^{\geq}) = \bar{C}(Cl_t^{\geq}) - \underline{C}(Cl_t^{\geq})$,
- $Bn(Cl_t^{\leq}) = \bar{C}(Cl_t^{\leq}) - \underline{C}(Cl_t^{\leq})$.

The boundaries group objects that can neither be ruled in nor out as members of class Cl_t.

In our example, there are four preference-ordered classes, namely: Cl_1, Cl_2, Cl_3 and Cl_4. Thus, the class unions that should be approximated are:

- $Cl_1^{\leq}$, i.e. compositions belonging to (at most) class Cl_1,
- $Cl_k^{\leq}$, i.e. compositions belonging to at most class Cl_k, $k = 2, 3$,
- $Cl_k^{\geq}$, i.e. compositions belonging to at least class Cl_k, $k = 2, 3$,
- $Cl_4^{\geq}$, i.e. compositions belonging to (at least) class Cl_4.

These class unions have been approximated using the equations given above. The result of the approximation is summarised in Table 3. As shown in this table, all the boundaries are empty sets, which indicates that the approximation is perfect.

Table 3. Approximations.

Class union	Lower approximation	Upper approximation	Boundary
$Cl_1^{\leqslant}$	5, 8	5, 8	$\emptyset$
$Cl_2^{\leqslant}$	2, 4, 5, 6, 7, 8	2, 4, 5, 6, 7, 8	$\emptyset$
$Cl_3^{\leqslant}$	1, 2, 3, 4, 5, 6, 7, 8	1, 2, 3, 4, 5, 6, 7, 8	$\emptyset$
$Cl_2^{\geqslant}$	1, 2, 3, 4, 6, 7, 9, 10, 11	1, 2, 3, 4, 6, 7, 9, 10, 11	$\emptyset$
$Cl_3^{\geqslant}$	1, 3, 9, 10, 11	1, 3, 9, 10, 11	$\emptyset$
$Cl_4^{\geqslant}$	9, 10, 11	9, 10, 11	$\emptyset$

The obtained approximation can be evaluated through its quality and accuracy. The *quality of classification* of a partition Cl is defined as follows:

$$\gamma(Cl) = \frac{|U - ((\bigcup_{t \in T} Bn(Cl_t^{\geqslant})) \bigcup (\bigcup_{t \in T} Bn(Cl_t^{\leqslant})))|}{|U|}. \tag{2}$$

The accuracy of the rough-set representation of unions of classes $Cl_t^{\geqslant}$ and $Cl_t^{\leqslant}$ are computed as follows:

$$\alpha(Cl_t^{\geqslant}) = \frac{\underline{C}(Cl_t^{\geqslant})}{\bar{C}(Cl_t^{\geqslant})}. \tag{3}$$

$$\alpha(Cl_t^{\leqslant}) = \frac{\underline{C}(Cl_t^{\leqslant})}{\bar{C}(Cl_t^{\leqslant})}. \tag{4}$$

The quality of classification and accuracy of decision classes are summarised in Table 4, which indicates a perfect classification and accuracy.

Table 4. Quality of classification and accuracy.

Quality of classification	Accuracy					
	$Cl_1^{\leqslant}$	$Cl_2^{\leqslant}$	$Cl_3^{\leqslant}$	$Cl_2^{\geqslant}$	$Cl_3^{\geqslant}$	$Cl_4^{\geqslant}$
1	1	1	1	1	1	1

An important output of the DRSA is attribute reducts. A *reduct* is a subset of condition attributes that can, by itself, fully characterise the knowledge in the decision table. A reduct is minimal (with respect to inclusion) subset of condition attributes in the sense that no attribute can be removed from the reduct without deteriorating the quality of classification. The intersection of all reducts is called *core*. In the considered illustrative example, the analysis with the DRSA shows that there are two reducts, namely {ResponseTime, Availability} and {ResponseTime, Cost, Security}. The core is then {ResponseTime}.

4.3 Induction, Validation and Exploitation of Decision Rules

A decision rule is a consequence relation relating a set of conditions (premise) and a conclusion (decision). Each elementary condition is built upon a single condition attribute while a conclusion is defined as an assignment to decision classes. Three types of decision rules may be considered in DRSA: (i) certain rules generated from lower approximations; (ii) possible rules generated from upper approximations; and (iii) approximate rules generated from boundary regions. The general structures of certain decision rules are as follows:

- **If** *condition(s)*, **then** *At Most* Cl_t
- **If** *condition(s)*, **then** *At Least* Cl_t

Decision rules induction is an NP-hard problem, therefore heuristics are naturally used for this purpose. The most popular rule induction algorithm for the DRSA is DOMLEM [8], which generates a minimal set of rules. The application this algorithm on the approximation in Table 3 leads to a minimal set of 9 certain decision rules, which are given in Table 5. A minimal set means a set of non-redundant rules that cover all the compositions in the learning set. The last column in Table 5 indicates the strength of decision rules. The description of these rules is straightforward.

Induced decision rules need to be validated. In this paper, the reclassification analysis, which consists in using the generated decision rules in order to reclassify the original decision objects, has been used to validate the induced decision rules. The result of the reclassification can be summarised through an $h \times h$ confusion matrix, where h is the number of decision classes. The intersection of a row and column indicates the number of original and possible assignments for the decision classes corresponding to the considered row and column. The confusion matrix for the learning set considered in this paper is given in Table 6. It shows a perfect match between the original assignments and those proposed by the decision rules.

Table 5. Decision rules.

#	Description	Strength (%)
Rule 1	**If** (ResponseTime $\geqslant$ 10.42) **then** (QoS at most 1)	100
Rule 2	**If** (ResponseTime $\geqslant$ 9.52) **then** (QoS at most 2)	66.67
Rule 3	**If** (Cost $\geqslant$ 3.38) **then** (QoS at most 2)	16.67
Rule 4	**If** (Availability $\leqslant$ 0.41817) & (Security $\leqslant$ 1) **then** (QoS at most 2)	16.67
Rule 5	**If** (Availability $\leqslant$ 0.56506) **then** (QoS at most 3)	100
Rule 6	**If** (Availability $\geqslant$ 0.57107) **then** (QoS at least 4)	100
Rule 7	**If** (Security $\geqslant$3) **then** (QoS at least 3)	20
Rule 8	**If** (Cost $\leqslant$ 2.48) & (ResponseTime $\leqslant$ 9.22) **then** (QoS at least 3)	20
Rule 9	**If** (ResponseTime $\leqslant$ 10.12) **then** (QoS at least 2)	100

Table 6. Confusion matrix.

		Possible			
	/	Cl_1	Cl_2	Cl_3	Cl_4
Original	Cl_1	2	0	0	0
	Cl_2	0	4	0	0
	Cl_3	0	0	2	0
	Cl_4	0	0	0	3

Once validated, decision rules can be used to classify unseen decision objects. In our example, the validated decision rules are then used to classify a set of 229 unseen compositions of which 15 have been assigned to the best QoS class. The description of these best composition instances is given in Table 7. We should mention that the indexes in the first column of Table 7 are not necessarily those used in Table 2. Then, the user should select one from the best instances for deployment. This is clearly beyond the scope of this paper.

Table 7. Best compositions.

#	Description	Response Time (q_1)	Availability (q_2)	Cost (q_3)	Security (q_4)
2	$\{s_1^1, s_2^1, s_3^1, s_4^1, s_5^1, s_6^2\}$	9.02	0.61487	2.92	1
6	$\{s_1^1, s_2^1, s_3^1, s_4^1, s_5^3, s_6^2\}$	8.42	0.66987	3.1	3
8	$\{s_1^1, s_2^1, s_3^1, s_4^1, s_5^4, s_6^2\}$	8.72	0.59794	3.34	3
10	$\{s_1^1, s_2^1, s_3^1, s_4^1, s_5^5, s_6^2\}$	9.02	0.63179	2.98	3
12	$\{s_1^1, s_2^1, s_3^2, s_4^1, s_5^1, s_6^2\}$	8.72	0.60763	3.02	1
16	$\{s_1^1, s_2^1, s_3^2, s_4^1, s_5^3, s_6^2\}$	8.12	0.66199	3.2	2
20	$\{s_1^1, s_2^1, s_3^2, s_4^1, s_5^5, s_6^2\}$	8.72	0.62436	3.08	2
36	$\{s_1^1, s_2^2, s_3^2, s_4^1, s_5^3, s_6^2\}$	8.12	0.59512	3.1	2
26	$\{s_1^1, s_2^2, s_3^1, s_4^1, s_5^3, s_6^2\}$	8.42	0.60220	3	2
62	$\{s_1^2, s_2^1, s_3^1, s_4^1, s_5^1, s_6^2\}$	8.72	0.59002	3.02	1
70	$\{s_1^2, s_2^1, s_3^1, s_4^1, s_5^5, s_6^2\}$	8.72	0.60626	3.08	3
72	$\{s_1^2, s_2^1, s_3^2, s_4^1, s_5^1, s_6^2\}$	8.42	0.58308	3.12	1
80	$\{s_1^2, s_2^1, s_3^2, s_4^1, s_5^5, s_6^2\}$	8.42	0.59913	3.18	2
86	$\{s_1^2, s_2^2, s_3^1, s_4^1, s_5^3, s_6^2\}$	8.12	0.57787	3.1	2
96	$\{s_1^2, s_2^2, s_3^2, s_4^1, s_5^3, s_6^2\}$	7.82	0.57107	3.2	2

5   Implementation and Evaluation

A prototype partially supporting the proposed approach has been implemented using jUDDI, which is an open source UDDI implementation. The jUDDI registry has been extended by adding a QoS classification component (QCC) composed of the following elements: (i) QoSInscription that can be used by customers looking to take into account the QoS attributes during service selection, (ii) LearningSet that contains a subset of composite web services that will be used as a learning set, and (iii) QoSDescription which provides the preference and data type parameters of QoS attributes.

The registry was implemented using Apache jUDDI. MySQL was used to implement the jUDDI databases. The UDDI4J is a Java class library that provides a set of APIs to interact with a jUDDI. These APIs are grouped into three categories: Iniquity, Publication and Security. The extended registry includes extensions to the UDDI4J Inquiry and Publication APIs set in order to manipulate the QoS related data.

The QCC supports two types of interaction: service request and service provider. The main steps for a service request are: (1) subscription to QCC, (2) subscription validation, (3) learning set specification, and (4) decision rules inference. Service providers can interact with the QCC in order to publish their services through the following steps: (1) sending the evaluation of participating services, (2) validation by the registry manager, (3) evaluation of possible compositions, and (4) classification of the considered compositions.

We evaluated the prototype with respect to processing time and quality of classification. The dataset used for evaluation, which have the same characteristics as the sample dataset used for illustration (same attributes and same number of classes), contains 240 compositions defined based on the composition graph in Fig. 1. Results indicate that the processing time increases almost linearly with respect to the number of compositions in the learning set. Results also show that the quality of classification increases linearly up to about 30 compositions in the learning set. Then, the quality of classification is stabilised around about 87%.

6   Conclusion

A QoS aware Web service composition approach proposed in [1,3] has been reviewed and extended in this paper. A special attention has been given to the classification phase where a typical machine learning method, namely the DRSA, has been used to assign the compositions into different QoS decision classes. The proposed extension has been illustrated using a didactic example. Furthermore, it has been validated through the development of a prototype and evaluated using a relatively large dataset. We should mention that some of used technologies (mainly UDDI and BPEL) are relatively off trend. The use of more recent technologies (including WSO2, which is a registry/repository alternative to UDDI, REST as an alternative to SOAP and BizTalk as an alternative to BPEL) will be considered in our future research.

There are also several directions for future research. A first topic is to enhance the composition approach by adding appropriate techniques to automatise the construction of the learning set in order to reduce the cognitive effort required from the user. The procedure proposed in [2,5] can be used as a start point. A second topic to address is to use of other versions of DRSA, such as the Variable Consistency Dominance-based Rough Set Approach (VC-DRSA) [8] and Stochastic DRSA [6]. The VC-DRSA is a variant of the DRSA that enables relaxation of the conditions for assigning decision objects to the lower approximations by accepting a limited proportion of negative examples, which is particularly useful for large datasets. The Stochastic DRSA relaxes the rough approximations in the DRSA by allowing inconsistencies to some degree. A third topic to investigate is related to the extension of the framework to support dynamic composition. The basic change concerns the construction and evaluation of the compositions.

A   Illustration of Partial Evaluation of Compositions

This appendix illustrates the calculation of partial evaluations of compositions using the formula given in Sect. 3.4. Let us consider the composition graph in Fig. 1 and assume that the specifications of QoS attributes used are those in Table 1.

In the following, we provide the formula for computing $v_j(x)$ $(j = 1, \cdots, 4)$ for the four QoS attributes in Table 1. These formula apply for non-leaf nodes, i.e., $x \in X_i$ such that $\Gamma^+(x) \neq \emptyset$. For leaf nodes, i.e. $x \in X_i$ such that $\Gamma^+(x) = \emptyset$, the partial evaluation on a QoS attribute q_j is simply $v_j(x) = q_j(x)$.

Response time (q_1). The response time of a non-leaf node x is computed as follows:

$$v_1(x) = q_1(x) + \max\{v_1(y) : y \in \Gamma^+(x)\} \tag{5}$$

$$v_1(x) = q_1(x) \quad + \sum_{y \in \Gamma^+(x)} \pi(x, y) \cdot v_1(y) \tag{6}$$

Equation (5) applies for the <flow> or the sequential BPEL constructors while Eq. (6) applies when the constructor <switch> is used. Here: Φ_1 is the sum and Ω is the max (for Eq. (5)) or the sum (for Eq. (6)).

Availability (q_2). The availability of a non-leaf node x is computed as follows:

$$v_2(x) = q_2(x) \quad \cdot \prod_{y \in \Gamma^+(x)} v_2(y) \tag{7}$$

$$v_2(x) = q_2(x) \quad \cdot \sum_{y \in \Gamma^+(x)} \pi(x, y) \cdot v_2(y) \tag{8}$$

Equation (7) applies for the `<flow>` BPEL or the sequential constructors while Eq. (8) applies when the constructor `<switch>` is used. Here: Φ_2 is the **product** and Ω is the **product** (for Eq. (7)) or the **sum** (for Eq. (8)).

Cost (q_3). The cost attribute of a non-leaf node x is computed as follows:

$$v_3(x) = q_3(x) \quad + \sum_{y \in \Gamma^+(x)} v_3(y) \tag{9}$$

$$v_3(x) = q_3(x) \quad + \sum_{y \in \Gamma^+(x)} \pi(x,y) \cdot v_3(y) \tag{10}$$

Equation (9) applies for the `<flow>` or the sequential BPEL constructors while Eq. (10) applies when the constructor `<switch>` is used. Here, the **sum** operator is used for both Φ_3 and Ω.

Security (q_4). The security of a non-leaf node x is computed as follows:

$$v_4(x) = \min\{q_4(x), \min_{y \in \Gamma^+(x)} \{v_4(y)\}\} \tag{11}$$

Here, both Φ_4 and Ω are the **min** operator. Recall that security attribute is an ordinal one. Equation (11) applies when the `<flow>` BPEL constructor is used. When the constructor `<switch>` is used, one of the rules mentioned above is used.

For illustration, we assume that the partial evaluations of providers of Web services S_1 to S_6 in Fig. 1 are as in Table 8. Supposing also that $p_1 = 0.4$ and $p_2 = 0.6$. Then, partial evaluation of composition k_{49} with respect to response time (q_1) is computed as follows. First, for leaf-node s_6^1, $v_1(s_6^1) = q_1(s_6^1) = 3.0$ holds. Then:

- Nodes s_4^1 and s_5^5: Eq. (5) leads to $v_1(s_4^1) = q_1(s_4^1) + \max\{v_1(s_6^1)\} = 4.8$ and $v_1(s_5^5) = q_1(s_5^5) + \max\{v_1(s_6^1)\} = 6.0$.
- Node s_3^1: Eq. (6) gives $v_1(s_3^1) = q_1(s_3^1) + (p_1 \cdot v_1(s_4^1) + p_2 \cdot v_1(s_5^5)) = 7.52$.
- Node s_2^3: Eq. (5) leads to $v_1(s_2^3) = q_1(s_2^3) + \max\{v_1(s_6^1)\} = 5.5$.
- Node s_1^1: Eq. (5) gives $v_1(s_1^1) = q_1(s_1^1) + \max\{v_1(s_2^3), v_1(s_3^1)\} = 9.52$.

The partial evaluation of composition k_{49} on QoS attribute response time is then $q_1(k_{49}) = 9.52$.

Table 8. Evaluations of individual Web services S_1 to S_6.

Service	Response Time (q_1)	Availability (q_2)	Cost (q_3)	Security (q_4)
s_1^1	2.0	0.99	0.4	1
s_1^2	1.7	0.95	0.5	2
s_1^3	1.6	0.80	0.7	3
s_1^4	3.4	0.94	0.1	5
s_2^1	2.0	0.99	0.7	5
s_2^2	3.0	0.89	0.6	2
s_2^3	2.5	0.82	0.4	1
s_3^1	2.0	0.85	0.4	4
s_3^2	1.7	0.84	0.5	2
s_4^1	1.8	0.89	0.3	5
s_5^1	3	0.86	0.5	1
s_5^2	1.5	0.60	0.6	2
s_5^3	2	0.99	0.8	5
s_5^4	2.5	0.82	1.2	4
s_5^5	3	0.90	0.6	3
s_6^1	3	0.8	0.23	5
s_6^2	2.5	0.92	0.5	3

References

1. Chakhar, S.: QoS-enhanced broker for composite Web service selection. In: Proceedings of the 8th International Conference on Signal Image Technology & Internet Based Systems, Sorrento-Naples, Italy, pp. 533–540, September 2012
2. Chakhar, S., Ishizaka, A., Thorpe, A., Cox, J., Nguyen, T., Ford, L.: Calculating the relative importance of condition attributes based on the characteristics of decision rules and attribute reducts: Application to crowdfunding. Eur. J. Oper. Res. **286**(2), 689–712 (2020)
3. Chakhar, S., Haddad, S., Mokdad, L., Mousseau, V., Youcef, S.: Multicriteria evaluation-based framework for composite web service selection. In: Bisdorff, R., Dias, L.C., Meyer, P., Mousseau, V., Pirlot, M. (eds.) Evaluation and Decision Models with Multiple Criteria. IHIS, pp. 167–200. Springer, Heidelberg (2015). https://doi.org/10.1007/978-3-662-46816-6_6
4. Chen, Y., Huang, J., Lin, C., Shen, X.: Multi-objective service composition with QoS dependencies. IEEE Trans. Cloud Comput. **7**(02), 537–552 (2019)
5. Dau, H.N., Chakhar, S., Ouelhadj, D., Abubahia, A.M.: Construction and refinement of preference ordered decision classes. In: Ju, Z., Yang, L., Yang, C., Gegov, A., Zhou, D. (eds.) UKCI 2019. AISC, vol. 1043, pp. 248–261. Springer, Cham (2020). https://doi.org/10.1007/978-3-030-29933-0_21

6. Dembczyński, K., Greco, S., Kotłowski, W., Słowiński, R.: Statistical model for rough set approach to multicriteria classification. In: Kok, J.N., Koronacki, J., Lopez de Mantaras, R., Matwin, S., Mladenič, D., Skowron, A. (eds.) PKDD 2007. LNCS (LNAI), vol. 4702, pp. 164–175. Springer, Heidelberg (2007). https://doi.org/10.1007/978-3-540-74976-9_18

7. Greco, S., Matarazzo, B., Słowiński, R.: Rough sets theory for multicriteria decision analysis. Eur. J. Oper. Res. **129**(1), 1–47 (2001)

8. Greco, S., Matarazzo, B., Slowinski, R., Stefanowski, J.: An algorithm for induction of decision rules consistent with the dominance principle. In: Ziarko, W., Yao, Y. (eds.) RSCTC 2000. LNCS (LNAI), vol. 2005, pp. 304–313. Springer, Heidelberg (2001). https://doi.org/10.1007/3-540-45554-X_37

9. Hu, C., Wu, X., Li, B.: A framework for trustworthy web service composition and optimization. IEEE Access **8**, 73508–73522 (2020)

10. Jatoth, C., Gangadharan, G., Fiore, U., Buyya, R.: QoS-aware big service composition using MapReduce based evolutionary algorithm with guided mutation. Futur. Gener. Comput. Syst. **86**, 1008–1018 (2018)

11. Liu, Y., Esseghir, M., Boulahia, L.M.: Cloud service selection based on rough set theory. In: The 2014 International Conference and Workshop on the Network of the Future (NOF), pp. 1–6 (2014)

12. Moghaddam, M., Davis, J.G.: Service selection in web service composition: a comparative review of existing approaches. In: Bouguettaya, A., Sheng, Q., Daniel, F. (eds.) Web Services Foundations, pp. 321–346. Springer, New York (2014). https://doi.org/10.1007/978-1-4614-7518-7_13

13. Pawlak, Z.: Rough Set. Theoretical Aspects of Reasoning About Data. Kluwer Academic Publishers, Dordrecht (1991)

14. Slowinski, R., Greco, S., Matarazzo, B.: Rough sets in decision making. In: Meyers, R. (ed.) Encyclopedia of Complexity and Systems Science, pp. 7753–7787. Springer, New York (2009). https://doi.org/10.1007/978-0-387-30440-3_460

15. Subbulakshmi, S., Ramar, K., Saji, A., Chandran, G.: Optimized web service composition using evolutionary computation techniques. In: Hemanth, J., Bestak, R., Chen, J.I.Z. (eds.) Intelligent Data Communication Technologies and Internet of Things. LNDECT, pp. 457–470. Springer, Singapore (2021). https://doi.org/10.1007/978-981-15-9509-7_38

16. Syu, Y., Kuo, S.M.J., FanJiang, Y.Y.: A survey on automated service composition methods and related techniques. In: Proceedings of the Ninth International Conference on Services Computing, Hawaii, USA, pp. 290–297, 24–29 June 2012

17. Yang, Y., et al.: ServeNet: a deep neural network for Web services classification. In: IEEE International Conference on Web Services (ICWS), pp. 168–175. IEEE Computer Society, Los Alamitos (2020)

18. Zhang, P., Zhang, Y., Dong, H., Jin, H.: Multivariate QoS monitoring in mobile edge computing based on Bayesian classifier and rough set. In: IEEE International Conference on Web Services (ICWS), pp. 189–196 (2020)

19. Zhao, X., Li, R., Zuo, X.: Advances on QoS-aware web service selection and composition with nature-inspired computing. CAAI Trans. Intell. Technol. **4**(3), 159–174 (2019)

Towards a Semantic Model
of the Context of Navigation

Federico Faruffini[1,2]([✉]) [iD], Alessandro Correa-Victorino[1] [iD],
and Marie-Hélène Abel[1] [iD]

[1] Université de Technologie de Compiègne, CNRS, Heudiasyc (Heuristics and
Diagnosis of Complex Systems), CS 60 319 - 60 203, Compiègne Cedex, France
federico.faruffini@etu.utc.fr,
{alessandro.victorino,marie-helene.abel}@hds.utc.fr
[2] Università degli Studi di Genova, Via Balbi, 5, 16126 Genoa, Italy
https://www.hds.utc.fr/
https://unige.it/en

Abstract. Many studies faced the problem of robot autonomous navigation in different fields, but nowadays just a few of them uses all the implicit information coming from the context in which such navigation is occurring. This results in a huge potential information loss that prevents us from adapting the robot's behaviour to each different situation it may be in. In this paper we therefore define the Context of Navigation using a semantic model built with ontologies. Then, we show what kind of inference is useful in such application. Finally, we provide a method to exploit the information coming from the context together with the control loop of the robot to better navigate it. Two application examples are provided to illustrate these concepts: a service robot and an intelligent autonomous vehicle.

Keywords: Decision support system · Navigation context modelling · Ontology · Robotics

1 Introduction

The field of robot autonomous navigation has met a great research interest since the Grand DARPA Challenges, held since 2004. Current state-of-the-art mathematical models are able to conduct global navigation autonomously, while performing a safe obstacle detection. Most of these robotic models are based on exteroceptive sensors to the robot, for instance cameras, lidars and other laser scanners, GPS systems and so on. A closely related field of research is the one of ADAS - Advanced Driving Assistance Systems - which are electronic systems built to help the driver while performing tasks as driving, overtaking, or parking.

Even if many of the existing solutions for autonomous navigation and ADAS systems show good results, a very few of them takes into account all the information coming from the semantic context in which the navigation is happening, resulting in a driving style which is not fully adaptable to the situation.

© Springer Nature Switzerland AG 2021
I. Saad et al. (Eds.): ICIKS 2021, LNBIP 425, pp. 168–183, 2021.
https://doi.org/10.1007/978-3-030-85977-0_13

Without taking into account the Context of Navigation, the autonomous vehicle misses many of the implicit information that usually a human driver knows, and to which adapts the way he drives. It is then clearly necessary to define a Context of Navigation and then to use the information it contains in order to change dynamically the way the robot/vehicle behaves in different situations. The application of such definition to different cases is complex as many information have to be taken into account, and also because we could have differences in the desired output of model: we can build a model more focused on reasoning or a more reactive one.

In a previous work [16] this problem was introduced, and a first theoretical approach was proposed for autonomous cars. In this work we introduce the modelling of the Context of Navigation, giving its definition and examples of its implementation. We also provide a second example to the one of the previous works, a service robot, to show how this method is applicable to any case of robotic navigation, through proper context modelling. Finally, we present the results of some tests we performed on the system.

In this work we study more in detail the modelling of the Context of Navigation, with a more complete result. We also provide a second example to the one of the previous works, a service robot, to show how this method is applicable to any case of robotic navigation, through proper context modelling.

This paper will proceed as follows: we are first going to propose some examples of problems we could solve with a semantic modelling of the context in Sect. 2, and then to give an overview of existing solutions both for autonomous navigation and driving assistance systems in Sect. 3. We then proceed to Sect. 4, which contains some of the reasons for the choice of such a model, while in Sect. 5 we provide with our definition of the Context and the different kinds of information we have to fit in it. Afterwards, in Sect. 6 we go through some of the different inference rules we can apply over our model. Our proposals for the interaction with the robot's control loop can be found in Sect. 7.

2   Problem Statement

In order to obtain a better adaptation of the robot's behaviour to the situation, we have to build a system to make it understand and exploit the contextual information during the navigation. In this paper we face the problem of enabling a robot to perform context-aware autonomous navigation. To do so, it is at first necessary to define the context in which such navigation occurs. Then, we have to define the way to reason over it and infer new information from it. Finally, we have to study a solution to interact with the robot's control loop in order to take into account the new information. We hereby present two situations of application of our work, to be used as example during this paper.

In the first example we have a service robot operating in an indoor home scenario. Its tasks are to operate many different home appliances in the house, as done by Nakamura et al. in [10], or to help the resident with other operations. This is a difficult task due to the environment and the many functions home appliances have. The navigation of the robot in the house must be safe for the humans and itself, and must preserve the integrity of the environment.

The second example is the following: we have an autonomous robotic vehicle, which may have one or more passengers on board. Such passengers may have preferences on the way the vehicle drives: for instance, they could like it to drive at a slower pace than the maximum one allowed by the traffic rules, or to avoid roads with speed bumps to protect the integrity of a fragile load. We want to build a system which allows the car to adapt its own behaviour to these preferences, without causing traffic issues in a real application.

3   Related Works

Many mathematical models exist to handle the safe global navigation of a robot, may it be a humanoid, a service one or an autonomous car. We now go through some existing solutions for navigation and for other robotic tasks which may benefit from the use of the semantic information given by the context of navigation.

In the field of mobile robots, Reinaldo et al. in [12] proposed a solution to model the robot navigation in unknown environments. In the study, the authors developed a system the robot can use to predict the future behaviour of the recognized objects - for instance a person - to make decisions on the navigation path. In the field of service robots, Sato et al. [13] proposed a solution to let an indoors robot pick up objects that are being pointed by a human. This task is made easier by the use of some contextual information. Luperto et al. [8] developed and tested a system to lead a service robot to navigate the home of an elderly resident to provide help. The tested tasks comprehend navigating the house to locate the user to assess his health state. One of the key problems of service robots is related to the correct object detection, since in a home scenario very often the target object is partially occluded by other objects. Lim et al. [6] proposed a structure to help the robot in this task by letting it consider the contextual information.

In the field of vehicle navigation, the autonomous model usually has a component called controller, which receives as input the desired position of the car and gives as output the actual controls (acceleration, braking, rotating the wheel etc.) to reduce as much as possible the error between the desired position and the actual one. There exist different approaches to this problem, for instance some are sensor-based, relying on sensors as GPS ones to obtain the current position, while others exploit the image given by the frontal camera. One of the latter solutions, proposed by Lima and Victorino in [7], exploits the images coming from the frontal camera and the laser scans to keep the vehicle in the center of the lane while proceeding forward. In the case an obstacle is perceived, the controller computes the best couple of angular and longitudinal velocities to overtake it. Even if this model is capable of performing the global navigation fully autonomously, it doesn't take into account any information besides the ones obtained by the endoceptive and exteroceptive sensors and a few others, as the speed limit, which is set to a static constant.

Other solutions tried to incorporate a few more external information on the car's behaviour, as in the study done by Regele in [11]. He proposed to decompose the topological structure of the road into simple graph arcs, which can be assigned labels for an easier decision-making by the intelligent vehicle. Such labels allow it to understand which are the points in which different lanes converge - which are positions with a higher risk of accidents - or which lanes are allowed or forbidden, and so on. This way it is possible to reduce the computational power required for the local path planning, by feeding the model with just the correct and useful input data, deleting the excess ones. Another solution, proposed by Schlenoff et al. in [14] aimed at building a navigation planner for robotic vehicles to take into account surrounding obstacles in order to make precise predictions on the outcomes of a collision with them, regarding damage to the passengers, load and the vehicle itself. With this predictions, the planner decides if it is necessary to avoid the obstacle or not.

In the field of Advanced Driving Assistance Systems, we can see some solutions which addressed the problem of using external information to aid the performance of the system. Armand et al. in [1] proposed a method to let a car exploit the perception of a pedestrian on the side of the road to predict its behaviour in order to have a quicker reaction time in the case breaking is needed. The authors explain how, with their model, a car may actively reason over the situation: in the case of a person close to a pedestrian crossing, for instance, the car understands she is likely to cross, and decreases its own speed. Some ADAS solutions aim to make the human driving safer, as in the case of Zhao et al.: the authors proposed semantic models to help the driver navigate uncontrolled intersections [19] and to prevent him from exceeding the speed limit [18].

Even if the studies hereby presented provide good solutions for the indoor navigation, automatic driving or ADAS systems, only in [1] the context in which the navigation is happening plays an important role. Some of the other studies presented do this just partly [8,11,13,18]. Of course, this leads to the loss of many potential information that could be used to better adapt the robot's behaviour to the situation. In this paper we present the Context of Navigation and make examples on its application to service robots and autonomous navigation. Before the definition of the Context, however, it is necessary to motivate the choice of technology we made, which is represented by ontologies.

4   Why to Use a Semantic Model for This Application

An ontology is a semantic structure to store real world data with a formal representation. The most important benefit in using an ontology structure is the possibility to exploit reasoners, which are pieces of software which can operate logical deductions on the asserted knowledge and data, and provide new information over it. For this reason, an ontology-based structure fits well in our problem, as in fact many of the proposed related works in Sect. 3 already used them [1,6,11,13,14,18,19]. Also, the context can be used to obtain a better interaction with a system of systems (in this case, an example could be that of many different vehicles on the same road): in [5] a context-aware recommender system for such a system was proposed.

An ontology-based structure works, in this situation, better than a DBMS-based one for different reasons besides the possibility of reasoning over them. For instance, they have a modular nature which allows the designer to update their structure after the detection. Also, they can be uploaded on the Web to be easily reused in other ontologies, helping in the development phase.

With a semantic model it is possible to store contextual information, as the age and preferences of each passenger of an intelligent car, the kind and fragility of the carried load, the conditions of the road, the environmental conditions and so on. However, it can still be used for storing essential physical data of our robot, as its max longitudinal and angular velocities, its suspensions performances, the kind of equipped tires, its length and so on. In order to define the context and the relative ontology we took inspiration from the methodology proposed in [14]. In this study, the authors give some suggestions on how to shape an ontology for autonomous navigation by asking some questions on the context and the extent to which we want our information to be precise. We then proceeded expanding the Context to fit all the information that are relevant to the navigation.

Some vehicle-related ontology-based standards already exist, and they are worth mentioning before moving on with the Context definition. The Vehicle Ontology[1] by Katsumi is proposed as a way to capture concepts related to vehicles. It contains many useful information related just to the vehicle and some of its data and object properties could be used in our scopes. For instance, for the first group we cite *accomodates bicycle, accomodates wheelchair, number of doors*, while for the second we have properties as *fuel consumption, fuel efficiency, drive wheel configuration*. Similar solutions can also be found in the Automotive Ontology Community Group[2] and in the Used Cars Ontology[3].

Our approach is different from the previous studies since it considers more information which are not taken into account in them. In fact, many of the ontologies proposed in the state of the art to solve the problem are somewhat similar to the ones just discussed, containing useful technical information, but just related to the vehicle, the topology of the road and/or the traffic rules. Our solution for the Context of Navigation aims at integrating this useful data with others which are missing in the previous works, as instance the ones related to the human users and their preferences.

---

[1] http://ontology.eil.utoronto.ca/icity/Vehicle/1.2/.

[2] https://www.w3.org/community/gao/.

[3] http://ontologies.makolab.com/uco/ns.html.

5   The Context of Navigation

In this section we illustrate our definition of the Context of Navigation and its structure. On the literature many definitions for Context have been developed, but most are too vague or too case-specific to our scopes. We found on Dey's one from [3] the most interesting:

> Context is any information that can be used to characterize the situation of an entity. An entity is a person, place, or object that is considered relevant to the interaction between a user and an application, including the user and applications themselves.

Following this definition, we then had to define what are the relevant entities to our application. In particular, we wanted to store information about the car, its passengers and load, the trip, the road and the other road users. Then, we gave our definition of Context of Navigation:

> The navigation context is any information that can be used to characterize the situation of navigation over a given period of time. Here, navigation is a movement considered relevant to the interaction between a driver and an application, including the driver and the applications themselves.

The Context of Navigation has two components: the Dynamic Context and the Static Context. The first contains all the information that vary during the trip with respect to the Navigation. For instance, for a service robot we could have the position of people or pets in the house, while in the example of vehicle navigation we could have traffic and right-of-way rules, the position of obstacles and their nature. The Static Context of Navigation contains information which don't change with respect to the navigation. In the case of a service robots we can make examples such as the shape of the rooms in the house or the different functions of the home appliances. In the case of intelligent navigation we can add to the Static Context information on the number of passengers, their driving preferences, the type of carried load, the model and performances of the car. Since the Dynamic Context of Navigation is more related to the path planning and obstacle avoidance, and many solutions for this already exist, we currently addressed mostly the Static Context.

Having decided to model our ontology in the OWL language with the Protégé editor [9], now we go through its main classes, data properties and object properties. To be noticed, to this day only a part of the full Context of Navigation has been implemented, but future works will complete it.

5.1   Context of Navigation in Service Robots

We now take the first example we defined in Sect. 2, about the service robot, and list some of the information about the Context of its navigation it may benefit from during its navigation. At the end of this subsection, we propose an example of the result of a first ontology with the information we discussed.

Person-Related Information. Since the robot will operate inside a home scenario, the people who live there have to be correctly encoded in the context. Of course, we need a class for *Person*. The residents have many attributes which are important to shape the context, as their age, their possible illnesses, the room they sleep in etc.etera. We then propose to model object and data properties as *hasAge, hasName, hasIllness*. We may also be interested in storing a state for each dweller, for example the one for sleeping, cooking and watching tv. This could be helpful in order, for example, not to wake up with noise an elderly person who is sleeping in the afternoon. Such a behaviour could be obtained by changing the speed of the displacement in order to reduce the noise, or by postponing non-urgent tasks. Also, when the information on the state of a person isn't available, we could shape a set of reasoning rules to let the robot infer in what state she could be, as we discuss in Sect. 6.

Object-Related Information. Of course, in order to obtain a well-defined Context of Navigation for a service robot it is of great importance to have a deep and precise structure for the surrounding objects. First of all, we have to define if a perceived object is recognize as static or not, for instance a chair will be and a person won't. This can be handled with a boolean property *isStatic*. Many static objects can be obstacles to the navigation, so this information is vital to the safety of the robot and the environment. Pets represent a kind of dynamic obstacle, which could be avoided implementing some kind of behaviour prediction as proposed in [12].

Besides being static or not, an object or obstacle could also represent a direct danger for the robot or the residents. For example, a robot must correctly perceive stairs as a really dangerous obstacle. We can do it with another object property, let's call it *isDangerous*. Other members of this category could be a pot on the stove or a newly washed floor, which could be temporary slippery.

An object could be marked as interactive, if such interaction is useful to the robot's tasks. For instance, all the house appliances are interactive and could be approached as proposed by studies like [10]. Also, obstacles like doors could be marked as interactive if the robot is allowed to use them to get into another room.

Action-Related Information. A service robot operating in a home scenario is supposed to perform many tasks different in their nature, as locating and reaching the user, bringing objects to her, operating the appliances and so on. It is therefore needed a way to encode the different actions into an ontology so that we can use it to reason and ameliorate the robot's behaviour. We need a class for *Action*, and the possibility to link its instances to form a plan, for instance with the object property *hasFollowingAction*. Even if the planning operations will be performed by a planner and not by our ontology, we can still use the latter to improve the final plan taking into account the context (Fig. 1).

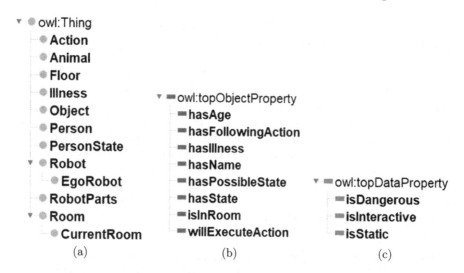

Fig. 1. The ontology for service robot navigation. (a) Classes; (b) Object Properties; (c) Data Properties

5.2 Context of Navigation in Autonomous Vehicles

We now see what are the different kinds of contextual information an intelligent vehicle can benefit from, along with our proposition on how to encode them into the Context.

Passenger-Related Information. There are many information related to the passengers that must be considered by the Context of Navigation. For sure, the passengers may have different preferences over the way the vehicle is driven, for instance some people may prefer a driving style that is slower and smoother than the standard one given by the automatic controllers, others may prefer to take specific paths even if they impact on the trip's duration. The comfort of a passenger is directly related to the derivative of the acceleration of the car, called the jerk, and this value could be different from person to person, while general suggestions and limit values already found [4]. With the definition of different preferences it will be possible to set a maximum (suggested) level of jerk for the car, for each trip. Another well-known problem in scientific literature is that of perceived safety in a robotic vehicle, since automatic models tend to drive in a different way from humans. Some studies propose to solve this issue instructing the model to execute more human-like paths, as in [17]. Again, this problem could be addressed with this type of Context. We can model OWL classes for the different situations in which we can have preferences, for instance for local roads, highways, interaction with specific objects and so on: each passenger will be able to set its own specific preference, that will be saved into an instance. We decided to have standard instances for such classes, so that if there is no preference we can drive the car normally, just following the traffic rules. An example for a local road is the class *LocalRoadDrivingStyle*, for which we have the standard instance *Standard_LocalRoadDrivingStyle* and the one by the passenger Emily, *Emily_LocalRoadDrivingStyle*.

The state of the passenger has to be taken into account too: a pregnant passenger may have specific needs, and a badly injured passenger needs that the vehicle takes him to the hospital as quickly as possible. Also, we could go further with some of these scenarios, for example the one of the pregnant woman: we may encode in the Context not only this fact, but all the other useful information regarding such pregnancy. For instance, we could use the information on the week of the pregnancy in order to understand if it is a safe period for travelling, the information on the healthyness of the mother, the one of the child and so on. The passengers could have preferences on the way to handle specific actions, as obstacle avoidance and overtaking.

Road-Related Information. Of course, in order to have a safe and legal driving style, it is necessary to encode information on the kind of road the vehicle is currently on. This will let it know the maximum allowed speed value, the possibility to switch to other lanes, the existing right-of-way rules etc. This can be enhanced by using road sign recognition systems, which are easily translatable into a semantic context. Country-specific regulations can be addressed with an approach similar to the modular one proposed by [2].

Load-Related Information. The Context should also model the information related to the load of the car for different reasons. For instance, for really heavy loads, we may want to let the robot know the approximate weight, so that it is capable of taking it into account in its reasoning and come up with a correct behaviour. This way, the car knows it may take longer to accelerate if compared to a scenario without load. Another information we could have regarding the carried load is its fragility: if a human driver knows she's carrying one or more fragile objects, she will adjust the way she drives and the path she chooses to reach the destination, for example avoiding sharp breakings or roads with speed bumps on them. It is enough to set a data property for encoding that information such that a robotic driver can drive accordingly, we can call it *isFragileObject*.

Robot-Related Information. As previously said, in order for the robot to make proper reasoning it is necessary to add to the ontology all the data that are related to its physical model. The level of the gas tank or of the battery, the number and kind of tires equipped, the maximum number of passengers and the current number on board, these are all examples of this kind of information. Even if it is possible to reason without them and still be able to modify the robot's behaviour, many of these information improve the quality of reasoning of the car, making it more adaptable but most importantly safer. For instance, if the car doesn't know if winter tires are equipped, if it starts snowing it can't take fully conscious decisions, possibly leading to dangerous situations. Also, the robot must be able to know if a spare tire is equipped, since it affects its stability at higher speeds, and set the maximum speed to a safer, lower value. Instead of using a simple boolean data property, in this case we could use an object property linking the car to a specific type of tire, with its own data.

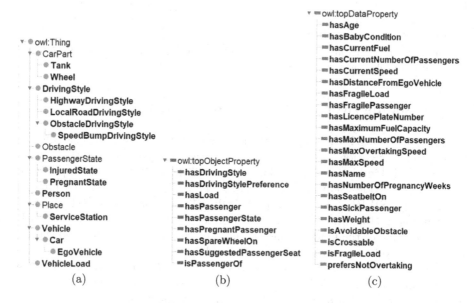

Fig. 2. The ontology for autonomous driving. (a) Classes; (b) Object Properties; (c) Data Properties

Obstacle Information. In order to enhance the tasks of safe navigation and obstacle avoidance, we have to model the different kinds of obstacles the robot could encounter in its path. If the robot has equipped an obstacle detection and recognition system, it is possible to try to label the external object and perform inference over it. In fact, the previously discussed ontology-based collision simulator proposed in [14] requires at least an estimation of the weight and material of the object to function. With this information it is possible to decide which action has the least bad consequences and to take it. Going further, we can try to assign different labels to known classes of obstacle: for instance, a traffic sign, a road hole or a tree could be labelled as static obstacles, while a car in front could be labelled as moving. Another example can be that of the perception of a child: this could be marked as an unpredictable obstacle, since even he is on the sidewalk, so he's not currently an obstacle, he may suddenly change its position and become one. A similar reasoning could be applied to the case of pets and wild animals.

Other Information. Of course our list is not exhaustive, since to model the whole context many more kinds of data are necessary. These comprehend for instance the weather conditions, the cyclic traffic behaviours, the presence of temporary strikes that close some roads, the time of the day and more. More complex reasoning rules can be implemented, for example by letting the vehicle know that, during the day hours, it has to drive slower in the roads that are adjacent to a school.

In Fig. 2 we can see the first version of the ontology we developed for autonomous navigation, following the steps above. Future works will aim to improve it to integrate more information.

6   Reasoning over the Context of Navigation

As previously stated, the main reason for the choice of ontologies to model the Context of Navigation is the possibility to perform reasoning over it. We now discuss which kind of rules our robot may need depending on the application and how to encode them into the Context. The rule language chosen for this application is SWRL - the Semantic Web Rule Language - and the chosen reasoner is Pellet [15], which is able to reason over OWL ontologies and SWRL rules too. Some of the kinds of rules we used in our ontology, in the application to intelligent vehicle navigation, are presented now.

Regarding the passenger's preferences we have different rules, as the ones directly impacting the maximum speed value. Let's take one of the examples we defined in Sect. 2, of a passenger who prefers to displace in a car which moves at a significantly slower speed than the maximum one set by the traffic rules. With an instance *Micheal* as such passenger, and *MyCar* as our ego vehicle, we can store the following triples:

$$MyCar \quad hasPassenger \quad Micheal$$
$$MyCar \quad hasLocalRoadDrivingStyle \quad Standard\_LocalRoadDrivingStyle$$
$$Standard\_LocalRoadDrivingStyle \quad hasMaxSpeed \quad 50$$

These three triples state that the car has *Micheal* as a passenger, and that the car is in an environment in which the maximum speed on local roads is 50 km/h.

$$Micheal \quad hasDrivingStylePreference \quad Micheal\_LocalRoadDrivingPreference$$
$$Micheal\_LocalRoadDrivingPreference \quad hasMaxSpeed \quad 30$$

With these triples we state that, when it comes to local roads, *Micheal* prefers to drive below 30 km/h. It is easy to create a very simple SWRL rule to adapt the maximum speed of the car to the one of the passenger, in the case there is only one person on board:

$$EgoVehicle(?v) \land hasPassenger(?v, ?p) \land Person(?p) \land$$
$$hasLocalRoadDrivingPreference(?p, ?pref) \rightarrow$$
$$hasLocalRoadDrivingStyle(?v, ?pref)$$

In the case of multiple passengers with different preferences, it is possible to define rules to choose just the lowest one, or the one of the pregnant passenger, or to use other criteria. Of course, in the case a passenger sets a preference for a speed over the one defined by the traffic rules, this won't be taken into account. To do this we can create a rule that, when different *hasMaxSpeed* values are present, always chooses the smallest.

More complex rules, defined over others can be defined, resulting in a multi-layered rule structure, as for the next two SWRL rules:

$$hasLoad(?v, ?l) \land VehicleLoad(?l) \land EgoVehicle(?v) \land isFragileLoad(?l, true) \rightarrow \\ hasFragileLoad(?v, true)$$

$$EgoVehicle(?v) \land hasFragileLoad(?v, true) \rightarrow \\ shouldAvoidRoadsWithObstacle(?v, SpeedBump)$$

This couple of rules will lead the car to choose, when possible, a path to its destination which doesn't present speed bumps in order not to damage the load. The safety of the trip may also depend on other information, as the presence of a mounted spare wheel, since these normally has a lower maximum speed than standard ones, a maximum number of kilometers to be used for, and other stats which are specific to the car and wheel's model and brand. We can model our ontology to store this information, and use SWRL rules to let the vehicle reason over it, by setting its own maximum speed with the one related to the kind of installed spare wheel.

Many other rules can be applied, for instance to modify the way the car overcomes speed bumps when a pregnant person is on board, in order not to hurt the child. Tests of this scenario were performed with good results.

Also, it is important to notice that many of these values for maximum speed must be intended as suggestions, as the ones of the preferences of the car. This means we don't expect the robotic vehicle to always follow them, but to try to adapt to these values when it is possible. However, there are cases in which it won't be possible or advisable for the vehicle to follow these suggestions. In Sect. 7 we give our solution to this problem.

7   Interaction with the Control Loop

In order to really adapt the behaviour of the car to the situation, just setting the maximum longitudinal speed as we proposed in Sect. 6 is not sufficient. To do so, we need to have a more complex interaction with the robot's control loop. For the sake of simplicity we consider for now just the longitudinal velocity of the robot and the case in which a passenger prefers the car to move at a slower speed than the standard one. A simple solution we propose is to add a block in the block diagram of the feedback control of the robot, which we called the Decision Block. The latter receives as inputs the desired velocities by the controller and the reasoner, and will output the actual one we want the robot to follow. In Fig. 3 we can see the new block diagram with the Decision Block. In this part of

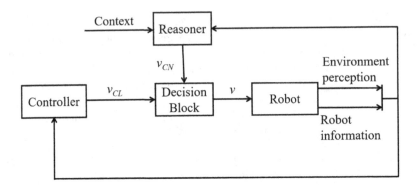

Fig. 3. The Decision Block outputs the reference speed for the vehicle to follow

the study, we proposed two ways the interaction between the ontology and the controller can be modelled. Since the first proposed solution was simply based on a modulation of the speed value given by the controller, we hereby present just the second one. We modelled our ontology, as we did in this paper, in such a way to output a *hasMaxSpeed* value for our vehicle, let's call it v_{CN} (as in velocity of Context of Navigation). The controller gives another reference output, let's call it v_{CL}. The mathematical rule for the Decision Block proposed is the following:

$$
v = \begin{cases} v_{CL}, & \text{if } v_{CL} \le v_{CN} \\ \gamma v_{CL} + (1 - \gamma) v_{CN}, & \text{otherwise} \end{cases} \tag{1}
$$

with $\gamma \in [0,1]$. (1) stands for the following: when the speed proposed by the controller is greater than the one suggested by the ontology, the final value of the reference velocity is computed through the convex combination of the two, with parameter γ. The value for this parameter has to be computed in real time, as it represents how much we follow the reasoner's suggestions. There could be cases, for instance one of little to no traffic, in which we can let the car proceed at a slow speed, so we can set $\gamma \cong 0$. On the other hand, in a really busy road this would create traffic issues and we can't let the car follow completely the suggestions. We then set $\gamma \cong 1$. Of course, values in-between may be used, for instance with $\gamma = 0.3$ we obtain a behaviour mostly based on the contextual suggestions, but with still a component related to the controller's output. In Fig. 4 we can see the velocity profile of a robotic car which longitudinal speed value was computed with Eq. 1. The example shows an overtaking scenario. The solid line represents the behaviour of the robot without context awareness, while the bottom one represents the behaviour when adapting to the speed preferences of the passenger. The lines in between represent the results we obtain with two different values of γ, as explained before.

Another solution is depicted in Fig. 5. Instead of using a Decision Block to merge the outputs of the controller and the reasoner, the reasoner output is fed to the controller as input. This means that the control law must be built or

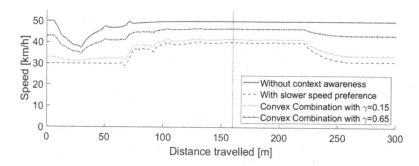

Fig. 4. Velocity profile of the robotic car applying (1)

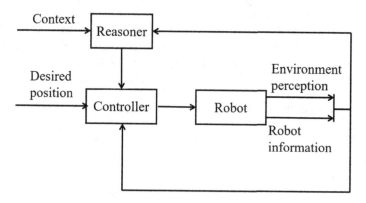

Fig. 5. The Reasoner's suggestions are taken into account directly in the robot's controller

modified accordingly to the way the ontology is shaped. We can also decide if to use the ontology as it is (for instance, suggesting a value of maximum speed as we did before) or a lighter version of it, maybe with simpler suggestions (*goSlower*, *goFaster* etc.). Also, since with this solution it will be possible to access the control loop, we can take into account values as the ones of the angular velocity, angular rate, jerk and others, and we can set specific rules on them.

8   Conclusions and Future Work

In this paper we showed how a semantic model of the Context of Navigation is required to let a robot adapt completely its behaviour to the contextual information at its disposal. In the use-cases of service robots and of autonomous vehicles, we showed how many contextual information can and must be considered to perform such a task. We have given our definition of the Context of Navigation and listed many of the areas of information that may be interesting in its scope. We have also already tested a first version of this solution in a simulator in two autonomous navigation scenarios, showing promising results. Tests on the service robot use-case will have to be performed too.

However, many more aspects of this work have to be deepened, starting with the full implementation of the ontology and rules of the Context of Navigation. Also, a deeper interconnection between it and the robot's control loop must be studied, in order to adapt the actual robot's control to the situation, instead of its output in speed. The problem of the performance of the reasoning system is of great importance in vehicle-related systems, as all the inference must be done in a matter of milliseconds for safety reasons. Many optimized reasoners were studied in the literature, and the most promising ones will have to be tested for this specific application.

Many questions remain open, as the way to provide the physical vehicle with the information: for instance, it could be unpleasant for the passengers to input their preferences each time they start the vehicle, and to solve this problem we could study a way to create a profile for each passenger. Of course, this will open security and ethical issues related to the personal data, that will have to be addressed.

Acknowledgement. This research study was funded by Idex Sorbonne Université - Labex MS2T.

References

1. Armand, A., Filliat, D., Ibanez-Guzman, J.: Ontology-based context awareness for driving assistance systems, pp. 227–233 (2014). https://doi.org/10.1109/IVS.2014.6856509
2. Buechel, M., Hinz, G., Ruehl, F., Schroth, H., Györi, C., Knoll, A.: Ontology-based traffic scene modeling, traffic regulations dependent situational awareness and decision-making for automated vehicles (2017). https://doi.org/10.1109/IVS.2017.7995917
3. Dey, A.: Understanding and using context. Pers. Ubiquitous Comput. **5**, 4–7 (2001). https://doi.org/10.1007/s007790170019
4. Elbanhawi, M., Simic, M., Jazar, R.: In the passenger seat: investigating ride comfort measures in autonomous cars. IEEE Intell. Transp. Syst. Mag. **7**(3), 4–17 (2015). https://doi.org/10.1109/MITS.2015.2405571
5. Li, S.: Context-aware recommender system for system of information systems. University of Technology of Compiègne (2021)
6. Lim, G.H., Suh, I.H., Suh, H.: Ontology-based unified robot knowledge for service robots in indoor environments. IEEE Trans. Syst. Man Cybern. - Part A: Syst. Humans **41**(3), 492–509 (2011). https://doi.org/10.1109/TSMCA.2010.2076404
7. Lima, D., Victorino, A.: A hybrid controller for vision-based navigation of autonomous vehicles in urban environments. IEEE Trans. Intell. Transp. Syst. 1–14 (2016). https://doi.org/10.1109/TITS.2016.2519329
8. Luperto, M., et al.: Towards long-term deployment of a mobile robot for at-home ambient assisted living of the elderly. In: 2019 European Conference on Mobile Robots (ECMR), pp. 1–6 (2019). https://doi.org/10.1109/ECMR.2019.8870924
9. Musen, M.A.: The protégé project: a look back and a look forward. AI Matters **1**(4), 4–12 (2015). https://doi.org/10.1145/2757001.2757003

10. Nakamura, T., Yuguchi, A., Maël, A., Garcia Ricardez, G.A., Takamatsu, J., Oga-sawara, T.: Ontology generation using GUI and simulation for service robots to operate home appliances. In: 2019 Third IEEE International Conference on Robotic Computing (IRC), pp. 315–320 (2019). https://doi.org/10.1109/IRC.2019.00058

11. Regele, R.: Using ontology-based traffic models for more efficient decision making of autonomous vehicles, pp. 94–99 (2008). https://doi.org/10.1109//ICAS.2008.10

12. Reinaldo, J.O., Maia, R.S., Souza, A.A.: Adaptive navigation for mobile robots with object recognition and ontologies. In: 2015 Brazilian Conference on Intelligent Systems (BRACIS), pp. 210–215 (2015). https://doi.org/10.1109/BRACIS.2015.50

13. Sato, E., Sakurai, S., Nakajima, A., Yoshida, Y., Yamguchi, T.: Context-based interaction using pointing movements recognition for an intelligent home service robot. In: RO-MAN 2007 - The 16th IEEE International Symposium on Robot and Human Interactive Communication, pp. 854–859 (2007). https://doi.org/10.1109/ROMAN.2007.4415204

14. Schlenoff, C., Balakirsky, S., Uschold, M., Provine, R., Smith, S.: Using ontologies to aid navigation planning in autonomous vehicles. Knowl. Eng. Rev. **18**(3), 243–255 (2003). https://doi.org/10.1017/S0269888904000050

15. Sirin, E., Parsia, B., Grau, B.C., Kalyanpur, A., Katz, Y.: Pellet: a practical OWL-DL reasoner. J. Web Semant. **5**(2), 51–53 (2007). https://doi.org/10.1016/j.websem.2007.03.004. https://www.sciencedirect.com/science/article/pii/S1570826807000169. Software Engineering and the Semantic Web

16. Victorino, A., Abel, M.H.: On the implementation of semantic model for intelligent vehicle navigation. In: Proceedings of the 2nd International Conference on Deep Learning, Artificial Intelligence and Robotics, ICDLAIR 2020 (2020)

17. Werling, M., Ziegler, J., Kammel, S., Thrun, S.: Optimal trajectory generation for dynamic street scenarios in a frenét frame. In: 2010 IEEE International Conference on Robotics and Automation, pp. 987–993 (2010). https://doi.org/10.1109/ROBOT.2010.5509799

18. Zhao, L., Ichise, R., Mita, S., Sasaki, Y.: An ontology-based intelligent speed adaptation system for autonomous cars. In: Supnithi, T., Yamaguchi, T., Pan, J.Z., Wuwongse, V., Buranarach, M. (eds.) JIST 2014. LNCS, vol. 8943, pp. 397–413. Springer, Cham (2015). https://doi.org/10.1007/978-3-319-15615-6_30

19. Zhao, L., Ichise, R., Yoshikawa, T., Naito, T., Kakinami, T., Sasaki, Y.: Ontology-based decision making on uncontrolled intersections and narrow roads. In: 2015 IEEE Intelligent Vehicles Symposium (2015). https://doi.org/10.1109/IVS.2015.7225667

Author Index

Printed in the United States
by Baker & Taylor Publisher Services